Manual *of* Immunological Methods

Pauline Brousseau
Yves Payette
Helen Tryphonas
Barry Blakley
Herman Boermans
Denis Flipo
Michel Fournier

With the Collaboration of
Martin Beaudet • *Edouard Kouassi*
Patrice Lapierre • *Isabelle Voccia*

Sponsored by
Canadian Network of Toxicology Centres

D1334843

CRC Press
Boca Raton Boston London New York Washington, D.C.

Library of Congress Cataloging-in-Publication Data

Manual of immunological methods / by P. Broussaeu ... [et al.] ;
   with the collaboration of M. Beaudet ... [et al.]
      p.   cm.
    Includes bibliographical references.
    ISBN 0-8493-8558-X (alk. paper)
     1. Immunology--Technique--Handbooks, manuals, etc.
  2. Immunoassay--Handbooks, manuals, etc. I. Brousseau, P.
(Pauline) II. Beaudet, M.
QR183.H36   1998
571.9'6'028--dc21   98-3373
                                           CIP

© 1999 by CRC Press LLC

No claim to original U.S. Government works
International Standard Book Number 0-8493-8558-X
Library of Congress Card Number 98-3373
Printed in the United States of America 1 2 3 4 5 6 7 8 9 0
Printed on acid-free paper

Table of Contents

Chapter 3
Removal of Organs

Chapter 4
Preparation of Cell Suspensions

Chapter 8
Oxidative Burst Assay Using Flow Cytometry

Chapter 9
Cell Cytotoxicity

Foreword

The Canadian Network of Toxicology Centres (CNTC) is a non-profit network of university-based collaborating scientists dedicated to research, training, and risk assessment and communication. We believe that effective improvements to environmental and human health can only be achieved in an objective research environment, supported by meaningful risk communication so that Canadians can better understand both risks and benefits *and* thereby make informed decisions based on factual information. In this regard, it has been our privilege to work with industry-based organizations, government agencies, and the international community in order to identify important gaps in our knowledge and to develop realistic research goals to address these critical gaps through the expertise of our multidisciplinary research teamwork approach.

The CNTC has provided funding support for immunotoxicology research of the Immunotoxicology Research Team for the past 5 years. It is my pleasure to acknowledge the efforts of the scientists across Canada who have been involved in the research, interpretation of data, and the preparation of the *Manual of Immunological Methods*.

Len Ritter

Preface

Since the last decade, immunology has become a very complex science that has found applications in a variety of fields. Although fundamental research is still needed because all of the mechanisms by which the immune system plays its crucial role in the maintenance of the homeostasis of organisms are still not yet elucidated, there is uniform agreement among scientists that the immune system, like other organ systems, could be the target of xenobiotic action. Consequently, new areas of research have appeared. Immunopharmacology, which leads to immunotherapy and which is defined as the preclinical and clinical science of immune manipulation and immunotoxicology, then stands as the field to understand how immunosuppression and immunotoxicity arise. The need to test for chemical-induced immunotoxic effects has led to the development and refinement of a plethora of quantitative and functional immunoassays.

However, it is well known that immunological tests, especially functional assays, could provide heterogeneous results leading to ambiguous conclusions to be used in efficacy study or hazard identification of immunotoxicity as well as in risk-assessment evaluation. A variety of reasons could explain this situation. Among them, factors such as age and genetic background of the subjects as well as the source of reagents, could interfere. Nevertheless, even if many of these variables could be made uniform, heterogeneity of results might be still a problem if the methodologies themselves are not standardized. Recently, this concern has been reinforced by two international harmonization efforts.

Our primary goal then is to sensitize people working in immunology, those from research institutes, universities, hospitals, and those from the private sector, to the fact that in order to enhance the successful application of this science, an important effort is needed to improve the quality of experiments performed and the consistency of results obtained.

In this volume, we provide an important number of assays that have been validated. For many of them, their validation was done through a harmonization program within the immunology program of the Canadian Network of Toxicology Centres. All of the assays are accompanied by a series of tools that will guide readers to perform their laboratory work in the spirit of the Good Laboratory Practices.

The document also includes a set of methods applicable to the study of environmental immunotoxicology in a variety of species more representative of wildlife — earthworms, bivalves, fish (freshwater and marine), birds, and mammals, including beluga whales and humans.

Acknowledgment

The authors would like to thank the Canadian Network of Toxicology Centres, the Department of Fisheries and Oceans at Maurice Lamontagne Institute, the TOXEN, the Canadian Wildlife Service, the U.S. EPA, and Health Canada for financial support.

We are grateful to the Université du Québec à Montréal, the University of Saskatchewan, the University of Guelph, Maisonneuve-Rosemont Hospital, the Zoo of Granby, and Shed Aquarium for providing samples.

Contributors

Martin Beaudet, M.Sc.
Université du Québec à Montréal
Centre TOXEN
C.P. 8888, Succ. Centre-Ville
Montreal, (Quebec)
H3C 3P8

Corinne Benquet, Ph.D.
Centre De recherche
Hôpital Maisonneuve-Rosemont
5415, Boulevard de l'Assomption
Montreal, (Quebec)
H1T 2M4

Barry R. Blakley, DVM, Ph.D.
Dept. of Veterinary Physiological
  Sciences
Western College of Veterinary Medicine
University of Saskatchewan
52 Campus Drive
Saskatoon, (Saskatchewan)
S7N 5B4

Herman Boermans, Ph.D.
Biomedical Sciences
University of Guelph
Guelph, (Ontario)
N1G 2W1

Pauline Brousseau, Ph.D.
Concordia University
1455, de Maisonneuve Blvd. W.
Montreal, (Quebec)
H3G 1M8

Catherine C. Coghlin, M.Sc.
Dept. of Veterinary Physiological
  Sciences
Western College of Veterinary Medicine
University of Saskatchewan
52 Campus Drive
Saskatoon, (Saskatchewan)
S7N 5B4

Guylaine Ducharme, B.Sc.
Université du Québec à Montréal
Centre TOXEN
C.P. 8888, Succ. Centre-Ville
Montreal, (Quebec)
H3C 3P8

Denis Flipo, M.Sc.
Université du Québec à Montréal
Centre TOXEN
C.P. 8888, Succ. Centre-Ville
Montreal, (Quebec)
H3C 3P8

Michel Fournier, Ph.D.
Université du Québec à Montréal
Centre TOXEN
C.P. 8888, Succ. Centre-Ville
Montreal, (Quebec)
H3C 3P8

Edouard Kouassi, Ph.D.
Centre de recherche
Hôpital Maisonneuve-Rosemont
5415, Boulevard de l'Assomption
Montreal, (Quebec)
H1T 2M4

Alexandra Lacroix, M.Sc.
Université du Québec à Montréal
Centre TOXEN
C.P. 8888, Succ. Centre-Ville
Montreal, (Quebec)
H3C 3P8

Patrice Lapierre, M.Sc.
Université du Québec à Montréal
Centre TOXEN
C.P. 8888, Succ. Centre-Ville
Montreal, (Quebec)
H3C 3P8

Ann Maslen
Biomedical Sciences
University of Guelph
Guelph, (Ontario)
N1G 2W1

Yves Payette, M.Sc.
Université du Québec à Montréal
Centre TOXEN
C.P. 8888, Succ. Centre-Ville
Montreal, (Quebec)
H3C 3P8

Judith E.G. Smits, Ph.D.
Dept. of Veterinary Pathology
Western College of Veterinary Medicine
University of Saskatchewan
52 Campus Drive
Saskatoon, (Saskatchewan)
S7N 5B4

Helen Tryphonas, Ph.D.
Department of Health
Toxicology Research Division, Food
  Directorate
Health Direction Branch, Bureau of
  Chemical Safety
Sir Frederic Banting Research Centre
  2202D1
Ross Avenue, Tunney Pasture
Ottawa, (Ontario)
K1A 0L2

Isabelle Voccia, Ph.D.
Université du Québec à Montréal
Centre TOXEN
C.P. 8888, Succ. Centre-Ville
Montreal, (Quebec)
H3C 3P8

Margaret Yole, B.Sc.
Dept. of Veterinary Physiological Sciences
Western College of Veterinary Medicine
University of Saskatchewan
52 Campus Drive
Saskatoon, (Saskatchewan)
S7N 5B4

List of Abbreviations

AO	Acridine Orange
ADCC	Antibody-Dependent Cytotoxic Cell
ANOVA	Analysis of Variance
BSA	Bovine Serum Albumin
CMFDA	5-Chloromethylfluorescein Diacetate
Con A	Concanavalin A
CPM	Counts per Minute
CRBC	Chicken Red Blood Cells
CTL	Cytotoxic T Lymphocytes
DCFH-DA	Dichlorofluorescin-Diacetate
DEAE	Diethylaminoethyl-Dextran
DHR 123	Dihydrorhodamine 123
DiO	3,3'-Dioctadecyloxacarbocyanine Perchlorate
DMSO	Dimethyl Sulfoxide
DNA	Deoxyribonucleic Acid
D-PBS	Dulbecco's Phosphate Buffer Saline
DPM	Disintegrations per Minute
EBSS	Earle's Balanced Salt Solution
EDTA	Ethylenediaminetetraacetic Acid
EGTA	Ethylene Glycol-bis(β-aminoethyl Ether)N,N,N',N'-Tetraacetic Acid
E:T	Effector:Target Ratio
FBS	Fetal Bovine Serum
FCS	Fetal Calf Serum
FITC	Isothyocyanate of Fluorescein
FL	Fluorescent Light
FMLP	N-Formyl-Methionyl-Leucyl-Phenylalanine
FSC	Forwardscatter
GAM	Goat Antimouse
GPC	Guinea Pig Complement
GSH	Glutathione
HBSS	Hanks' Balanced Salt Solution
HEPES	(N-[2-Hydroxyethyl]piperazine-N'-[2-Ethanesulfonic Acid])
IgG	Immunoglobulin G
IL-2	Interleukin-2
i.m.	intramuscular
i.p.	intraperitoneal
LAK	Lymphokine-Activated Killer
LBSS	Lumbricus Balanced Salt Solution
LGL	Large Granular Lymphocyte
LPS	Lipopolysaccharides

LU	Lytic Unit
M	Molar
MFI	Mean Fluorescence Intensities
MHC	Major Histocompatibility Complex
MLR	Mixed Lymphocyte Reaction
NaN$_3$	Sodium Azide
NEM	N-ethylmaleimide
NH$_4$Cl	Ammonium Chloride
NK	Natural Killer
OZ	Opsonized Zymosan
PBS	Phosphate Buffer Saline
PC	Percent Cytotoxicity
PFC	Plaque Forming Cell
PHA	Phytohemagglutinin
PI	Propidium Iodide
PMA	Phorbol Myristate Acetate
PMN	Polymorpho Nuclear
PWM	Pokeweed Mitogen
RBC	Red Blood Cells
ROS	Reactive Oxygen Species
SI	Stimulation Index
SLS	Sodium Lauryl Sulfate
SRBC	Sheep Red Blood Cells
SSC	Sidescatter
v/v	Volume to Volume Ratio
w/v	Weight to Volume Ratio

List of Species Included in the Handbook

Alligator (*Alligator mississippiensis*)
Beluga (*Delphinapterus leucas*)
Bovine (*Bos taurus*)
Cat (*Felis catus*)
Chicken (*Gallus domesticus*)
Dog (*Canis familiaris*)
Duck (*Aythya affinis*)
Earthworm (*Lumbricus terrestris*)
Guinea pig (*Cavia porca*)
Horse (*Equus cabalus*)
Human (*Homo sapiens*)
Mice (*Mus musculus*)

Mink (*Mustela vison*)
Mollusks (*Mya arenaria*)
Monkey (*Macacus fascicularis*)
Perch (*Perca flavescens*)
Pig (*Sus crofa*)
Polar bear (*Ursus maritimus*)
Rabbit (*Oryctolagus cuniculus*)
Rat (*Rattus norvegicus*)
Seal (*Phoca vitulina*)
Sheep (*Ovis aries*)
Trout (*Oncorhynchus mykiss*)

1 Identification, Anesthesia, and Euthanasia

1.1 INTRODUCTION

Animals used in research studies might come from a certified breeding facility, if a specific genotype or phenotype is required, or from the wild. If the study protocol requires data acquisition from individual animals, it is necessary to identify the animals.

The methods described for identification of laboratory animals are the ear notching used for rats and mice and the tagging of fish. Depending on the sampling needs, the animal may be anesthetized first or directly euthanized before removing the desired body fluids or organs. In the case of anesthesia, it should be performed by qualified personnel. Both procedures are performed with the following suggested materials and equipment.

1.2 IDENTIFICATION OF LABORATORY RODENTS BY EAR NOTCHING

1.2.1 REAGENT

1. 70% isopropanol

1.2.2 MATERIALS AND EQUIPMENT

1. Ear punch

1.2.3 PROCEDURE

1. Clean the punch with isopropanol before and after use.
2. Using the ear punch, punch the hole or notch on the animal's ear that will correspond to the number of the animal. The holes must be carefully placed so that all punch marks are easily discerned by the animal handler. Refer to Figure 1.1 for an example of an identification code.
3. Determine the sex of the animal and record it with its corresponding number.
4. If the body weight is required by the study protocol, record it with the corresponding number of the animal.
5. Continue until all animals have been processed.

Note: There is more than one way to identify laboratory animals. They can be tattooed (on the tail or the ears) or marked with a marker. However, when using a marker, the animals should be checked on a daily basis to make sure that the numbers are clearly visible.

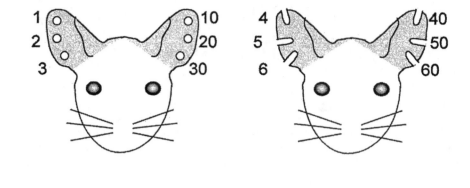

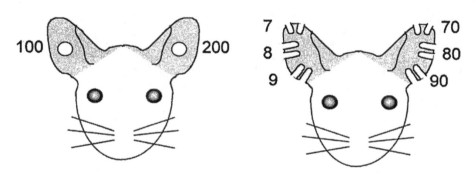

FIGURE 1.1   Identification of rodents.

1.3   IDENTIFICATION OF FISH BY FIN CLIPPING

There are several methods for marking fish, and the method of choice will depend on the species, age, size of fish, number of fish, cost, effects of the mark or tag on the fish, and the length of time the mark is required.

1.3.1   REAGENT

   1. 70% isopropanol

1.3.2   MATERIALS AND EQUIPMENT

   1. Small scissors
   2. Fish net
   3. Pan with wet towels

1.3.3 PROCEDURE

1. Clean scissors with isopropanol.
2. Fish must be anesthetized (see Section 1.4).
3. Remove the fish with a net taking care not to damage the integument.
4. Clip the adipose, the pelvic, or the pectoral fin and record animal number.

Note: Fins which are not clipped closely to the body surface may regenerate, and one may risk losing the mark. But if clipped too close, the fish may bleed and infection may occur.

1.4 ANESTHESIA OF FISH AND RODENTS

The type and concentrations of anesthetic vary with species and weight. Information on the spectrum of drugs and techniques is available in the Suggested Reading at the end of the chapter.

As an example, for the rat, we recommend the use of ketamine at 87 mg/kg in a mixture with xylazine at 7 mg/kg given i.p. or i.m., with a preanesthetic like acepromazine given i.p. or i.m. at a dose of 2.5 mg/kg. For the fish, tricaine methanesulfonate at a concentration of 60 to 100 mg/l diluted into a tank of water is commonly used with salmonids.

1.5 EUTHANASIA OF RODENTS

CO_2 provides an efficient method of euthanasia for mice and rats.

1.5.1 MATERIALS AND EQUIPMENT

1. Carbon dioxide (CO_2)
2. Desiccator jar

1.5.2 PROCEDURE

1. Place the animal at the bottom of the desiccator, close the lid, and attach the tubing to the gas cylinder.
2. Open the pressure valves to allow the CO_2 to flow at low pressure into the desiccator.
3. Before removing the animal, close all the valves.

Note: The animal should be exposed to a slowly rising concentration of the gas. This will ensure that the animal loses consciousness rather than suffocating as a result of immediate exposure to 100% CO_2. The exposure must last for at least 3 to 4 min.

1.6   EUTHANASIA OF FISH

1.6.1   REAGENT

   1. Anesthetic (i.e., tricaine methanesulfonate)

1.6.2   MATERIALS AND EQUIPMENT

   1. Tank filled with water
   2. Stick
   3. Fish net

1.6.3   PROCEDURE

   1. In the laboratory setting, transfer the fish with the net into a tank of water containing an overdose of anesthetic.
   2. In field conditions, strike the fish with a sharp blow on the head, which preferably will be preceded by slight anesthesia.

1.7 WORKING SHEET: WEIGHT OF LABORATORY ANIMALS

Date: _____   Species: _____

Experiment: _____   Prepared by: _____

Animal number	Weight	Treatment

Page ____ of ____

SUGGESTED READING

Harkness, J.E. and Wagner, Y.E., 1995. *The Biology and Medicine of Rabbits and Rodents,* 4th ed., Williams and Wilkins, Baltimore.

Iwama, G.K., 1992, Anesthesia, analgesia and euthanasia in fish, in *The Care and Use of Amphibians, Reptiles and Fish in Research, Proceedings from a SCAW/LSUSVVM–Sponsored Conference,* D.O. Schaeffer, K.M. Kleinow, L. Krulisch, Eds., April 8–9, 1991, New Orleans.

Waynforth, H.B. and Flecknell, P.A., 1994, *Experimental and Surgical Technique in the Rat,* 2nd ed., Academic Press, New York.

Whitman, K., 1996. Canadian Association for Laboratory Animal Science Aquatic Wet Lab, *Restraint, Handling and Clinical Techniques of Fish,* Atlantic Veterinary College, Halifax, July 8.

2 Collection of Peripheral Blood Samples

2.1 INTRODUCTION

The procedure described in this chapter deals with the collection of blood in different vertebrate species. Depending on the amount of blood needed and the type of study done, the animal may be exsanguinated or a small sample may be taken at different time intervals, leaving the animal alive for the duration of the study. The use of heparin or other anticoagulant is required to obtain plasma, while for serum collection no anticoagulant is needed.

Depending on the species and the experimental needs, blood collection can be performed via the orbital sinus, the caudal vein, the marginal ear vein, or by puncture. Whether or not the animal is to be sacrificed will influence the volume of blood that can be collected (see Table 2.1).

2.2 REAGENTS

1. Anesthetic*
2. 70% isopropanol
3. Anticoagulant (heparin,** EDTA, sodium citrate) or commercially available blood collection tubes
4. Petroleum jelly (for guinea pigs)

2.3 MATERIALS AND EQUIPMENT

1. Warming light (to induce vasodilatation of the veins)
2. Shaver
3. Hypodermic needles
4. Syringes
5. Aliquot mixer
6. Blood tubes
7. Capillary tubes cut to a length of 50 mm maximum (orbital sinus)

---

* Injectable anesthetics are usually effective, inexpensive, and convenient. Most are based on ketamine plus one or more drugs. The different types of anesthetic and their dosages may vary with species, weight, sex, age. See Suggested Reading at the end of Chapter 1.

** To perform a functional assay such as phagocytosis the use of heparin to avoid loss of calcium is recommended.

8. Hematocrit tubes (for small fish)
9. Shaker

2.4 PROCEDURE

2.4.1 CARDIAC PUNCTURE

This technique is useful in smaller species without accessible sites for collecting large volumes of blood (e.g., mouse, rat, chicken, fish, mink).

1. Anesthetize the animals and check vital signs such as respiration and heart rate.
2. Spray the thorax with alcohol and locate the heart by palpation
3. Insert the needle through the rib cage and into the heart until blood begins to fill the syringe; then draw back the plunger slowly until the necessary volume is collected.
4. Transfer blood into an appropriate tube, shake five times, and place on a shaker or in a test tube rack.

Cardiac puncture is a risky method for blood sampling and is best avoided if alternative methods are available.

Note: In fish, hold the needle perpendicular to the skin and insert it slightly below the tip of the V-shaped notch formed by the gill cover and the isthmus.

2.4.2 ORBITAL SINUS BLOOD COLLECTION

Used mainly on small rodents and rabbits.

1. Lightly anesthetize the animal.
2. Introduce the uncut end of the capillary into the normal nictitating membrane just lateral to the medial aspect of the eye with a full twisting motion.
3. Collect the blood into an appropriate container.
4. When finished, remove the capillary tube from the eye and place a gauze pad firmly against the eye to stop the bleeding.
5. The animal must be observed several times after the blood collection to prevent hemorrhage.
6. Transfer the blood into an appropriate tube, shake five times, and place on a shaker or in a test tube rack.

2.4.3 MARGINAL EAR VEIN BLOOD COLLECTION

Used mainly on rabbits and guinea pigs.

1. Restrain the animal, then shave the hair over the vein and swab with alcohol.
2. Dilate vein with a warming light.

3. Insert needle parallel to the vein and slowly draw back the plunger to fill the syringe with an appropriate volume of blood.
4. Apply pressure for approximately 20 s to the puncture site to stop the flow of blood.
5. Transfer the blood into an appropriate tube, shake five times, and place on a shaker or in a test tube rack.

2.4.4 CAUDAL TAIL VEIN BLOOD COLLECTION

Used in rat, mouse, fish, and beluga whales.

1. Dilate the caudal tail vein with a warming light; then swab the area with alcohol.
2. Insert the needle parallel to the vein and slowly draw back the plunger to fill the syringe with an appropriate volume of blood.
3. Apply pressure for approximately 20 s to the puncture site to stop the flow of blood.
4. Transfer the blood into an appropriate tube, shake five times, and place on a shaker or in a test tube rack.

Note:

- Because the vertebrae do not extend to the tip of a mouse tail, cutting a mouse tail (twice only) does not cut bone, as may occur in the rat.
- In fish, insert a needle under the scale of the ventral midline of the caudal peduncle of a freshly euthanized fish. Ease the needle toward the vertebral column until you reach the column and withdraw the needle a fraction of a millimeter until blood flows into the syringe. For small fish, sever the caudal peduncle and fill a hematocrit tube with the blood.

2.4.5 FEMORAL VEIN BLOOD COLLECTION IN MONKEYS

For monkeys, immobilize the animal on its back.

1. Dilate the femoral vein with a warming light; then swab the area with alcohol.
2. Insert the needle parallel to the vein and slowly draw back the plunger to fill the syringe with an appropriate volume of blood.
3. Apply pressure for approximately 20 s to the puncture site to stop the flow of blood.
4. Transfer the blood into an appropriate tube, shake five times, and place on a shaker or in a test tube rack.

2.4.6 CEPHALIC OR SAPHENOUS VEIN BLOOD COLLECTION IN DOGS, BEARS

Lay the animal on a table, pulling gently on its foreleg (cephalic) or hind leg (saphenous).

TABLE 2.1
Table of Blood Volumes Which Can Be Collected
from Different Species According to Body Weight

Species	Total Blood Volume (ml/kg)	% of Body Weight Containing Blood	Plasma Volume (ml/kg)
Cat	45–75	5.8	35–52
Dog	75–100	8.6	46–55
Fish	—	2.5–6.0 total body fluid	—
Guinea pig	65–90	7.2	35–48
Monkey	50–60	5.5	36–48
Mouse	70–80	7.0	40–50
Pig	55–65	6.0	35–39
Rabbit	45–70	5.7	28–51
Rat	50–65	6.4	36–45

1. Dilate the selected vein with a warming light; then swab the area with alcohol.
2. Insert the needle parallel to the vein and slowly draw back the plunger to fill the syringe with an appropriate volume of blood.
3. Apply pressure for approximately 20 s to the puncture site to stop the flow of blood.
4. Transfer blood into an appropriate tube, shake five times, and place on a shaker or in a test tube rack.

Note: Cephalic vein blood collection is more easily done than saphenous blood collection.

2.4.7 BRACHIAL VEIN BLOOD COLLECTION IN BIRDS

Lay the bird on a table. One person should hold the head and the legs while another strectches the wing.

1. Dilate the brachial vein with a warming light; then swab the area with alcohol.
2. Insert the needle parallel to the vein and slowly draw back the plunger to fill the syringe with an appropriate volume of blood.
3. Apply pressure for approximately 20 s to the puncture site to stop the flow of blood.
4. Transfer blood into an appropriate tube, shake five times, and place on a shaker or in a test tube rack.

2.4.8 JUGULAR VEIN BLOOD COLLECTION

Used successfully in larger animals such as dogs, sheep, mink, pig, bovine, and cat.

For the dog, sheep, pig, and bovine, restrain the animal. For the mink, anesthetize the animal sligthly. Clean the area with isopropanol and gauze pads.

1. Insert the needle parallel to the vein and slowly draw back the plunger to fill the syringe with an appropriate volume of blood.
2. Apply pressure for approximately 20 s to the puncture site to stop the flow of blood.
3. Transfer blood into an appropriate tube, shake five times, and place on a shaker or in a test tube rack.

SUGGESTED READING

Fletch, S.M. and Wobeser, G., 1970. A technique for safe multiple bleeding or intravenous injections in mink, *Can. Vet. J.,* 11, 33.

Flicknell, P.A., 1987. *Laboratory Animal Anaesthesia,* Academic Press, London.

Harkness, J.E. and Wagner, J.E., 1995. *The Biology and Medicine of Rabbits and Rodents,* 4th ed., Williams and Wilkins, Baltimore.

Hillyer, E.V., 1992. Blood collection techniques and normal clinical values in ferrets, *J. Small Exotic Anim. Med.,* 1, 178–180.

Waynforth, H.B. and Flecknell, P.A., 1992. *Experimental and Surgical Technique in the Rat,* 2nd ed., Academic Press, New York.

Whitman, K., 1996. Canadian Association for Laboratory Animal Science Aquatic Wet Lab, *Restraint, Handling and Clinical Techniques of Fish,* Atlantic Veterinary College, Halifax, July 8.

3 Removal of Organs

3.1 INTRODUCTION

For the collection of leukocytes, the tissues or organs associated with the immune system are a good source of cells and may be removed whether the animal has been bled or not and/or under sterile or nonsterile conditions. Sterile conditions demand extra care to prevent contamination of the organ and, subsequently, of the cell suspension.

In this chapter, we will describe the removal of the spleen, the thymus, the lymph node, and the bone marrow in the rat or mouse, as well as the removal of the thymus, spleen, and head kidney in the trout.

3.2 REAGENTS

1. Hanks' balanced salt solution (HBSS) containing 10 mM HEPES (sterile, if necessary)
2. HBSS containing 10 mM HEPES and 10 U/ml of heparin (HBSS complete medium)
3. 70% isopropanol
4. Anesthetic solution to anesthetize the fish (i.e., tricaine methanesulfonate)

3.3 MATERIALS AND EQUIPMENT

1. Tissue forceps (sterile, if necessary)
2. Scissors (sterile, if necessary)
3. Petri dish, 60 × 15 mm (sterile, if necessary)
4. Laminar flow cabinet
5. CO_2

3.4 PROCEDURE

Note: When sterile conditions are required, all steps after the death of the animal must be performed in a laminar flow hood.

3.4.1 REMOVAL OF THE SPLEEN

3.4.1.1 Rat or Mouse (see Figure 3.1)

1. Euthanize the animal with CO_2 anoxia (see Chapter 1) and soak with isopropanol.
2. Lay the animal on its right side. With tissue forceps, pinch and lift the skin of rat or mouse and cut it with scissors. Make an incision of 2 cm below the ribs without piercing the peritoneum to see the spleen clearly.

3. Lift the peritoneum and make a small incision to expose the spleen. Lift the spleen high and cut to release it.
4. Place the spleen in a petri dish and immerse in sterile HBSS.

3.4.1.2 Trout Model (see Figure 3.2)

1. Euthanize the animal (see Chapter 1) and then soak in isopropanol.
2. Lay the trout on its right side. With scissors, cut the skin from the anus to the head.
3. With forceps, lift the skin to expose the spleen; then lift it and cut to release the spleen.
4. Place the spleen in a petri dish and immerse in sterile HBSS complete medium.

3.4.2 REMOVAL OF THE THYMUS

3.4.2.1 Rat or Mouse

1. Euthanize the animal by CO_2 anoxia (see Chapter 1) and soak with isopropanol.
2. Lay the animal on its back. With forceps, pinch and lift the skin on the ribs of rat or mouse, then cut it with scissors.
3. Cut the skin under the peritoneum; lift the rib cage by cutting the ribs. The thymus will appear just above the heart.
4. Place the thymus in a petri dish and immerse in sterile HBSS.

3.4.2.2 Trout

1. Euthanize the animal (see Chapter 1).
2. Lay the animal on its side. With forceps, lift the operculum to expose the thymus, which is at the upper junction of the operculum and the gill. Using a small pair of scissors, detach the thymus and place it in a petri dish containing sterile HBSS complete medium.
3. To remove the other thymus, place the trout on the other side and repeat step 2.

3.4.3 REMOVAL OF THE LYMPH NODE (MOUSE OR RAT)

1. Euthanize the animal by CO_2 anoxia and soak with isopropanol.
2. Depending on the node to be taken, lay the animal on its right or left side. With forceps, pinch and lift the skin of rat or mouse; then cut it with scissors.
3. The lymph nodes can be found directly under the skin. Detach the lymph nodes and place in a petri dish with sterile HBSS.

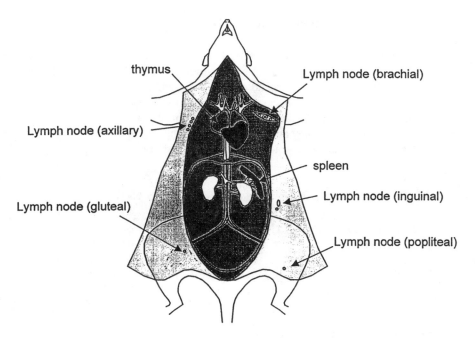

FIGURE 3.1 Lymphoid organs in rat.

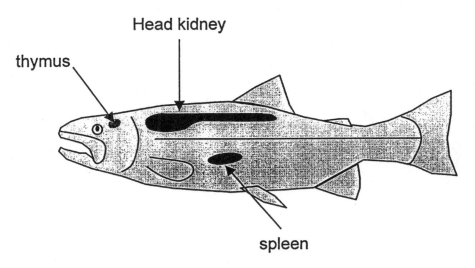

FIGURE 3.2 Lymphoid organs in fish.

3.4.4   REMOVAL OF THE HEAD KIDNEY (TROUT)

1. Euthanize the animal (see Chapter 1).
2. Lay the trout on its side. With scissors cut the skin from the anus to the pectoral fins; then cut perpendicularly to the top of the head.
3. With forceps, pull the viscera gently to expose the head kidney. Remove the swim bladder and the mesentery overlying the kidney.
4. Locate the main kidney and trace forward to the region of the head kidney. Excise the head kidney and place it in a petri dish with sterile HBSS complete medium.

3.4.5   REMOVAL OF THE BONE MARROW (MOUSE AND RAT)

1. Euthanize the animal by CO_2 anoxia and soak with isopropanol.
2. Remove both femurs, wash them in HBSS; then cut the epiphyses at each end.
3. Using a syringe filled with HBSS, flush the medium through the bone to remove the bone marrow. Collect the marrow in a petri dish with HBSS.

SUGGESTED READING

Feldman, D.B. and Seeley, J.C., 1988. *Necropsy Guide: Rodents and the Rabbit,* CRC Press, Boca Raton, FL, 167 pp.
Stoskopf, M.K., 1993. *Fish Medicine,* W.B. Saunders, Philadelphia, 882 pp.
Waynforth, H.B. and Flecknell, P.A., 1992. *Experimental and Surgical Techniques in the Rat,* 2nd ed., Academic Press, San Diego, 382 pp.

4 Preparation of Cell Suspensions

4.1 INTRODUCTION

To perform assays it is crucial to have cell suspensions of high quality with a viability greater than 80%. This chapter describes the preparation of cell suspensions from the peripheral blood, peritoneal exudate, and from organs of vertebrates, such as the duck, the rat or mouse, and the trout, and of invertebrates, such as the earthworm and mollusks.

The cells are collected from the thymus, spleen, and the lymph node in the mouse or rat and the thymus, spleen, and the head kidney in the trout. The removal of these organs is described in Chapter 3. Blood collection is described in Chapter 2. Cells in the earthworm are obtained by extrusion and in mollusks by hemolymph puncture.

4.2 EXTRUSION OF COELOMOCYTES IN THE EARTHWORM

4.2.1 REAGENTS

1. Saline solution: 0.85% NaCl (w/v) in deionized, distilled water, kept at +10°C
2. Extrusion solution: 0.85% saline solution, 5% ethanol, 3.5 mg/ml tetrasodium EDTA, and 10 mg/ml guaiacol glyceryl ether, adjusted to pH 7.3
3. Lumbricus balanced salt solution (LBSS) with calcium: 143 mM NaCl + 4.8 mM KCl + 3.8 mM $CaCl_2$ + 1.1 mM $MgSO_4 \cdot 7H_2O$ + 0.4 mM KH_2PO_4 + 0.3 mM $Na_2HPO_4 \cdot 7H_2O$ + 4.2 mM $NaHCO_3$; pH adjusted to 7.3 (osmolarity should be at 300 mOsM)
4. LBSS without calcium: same as above but without $CaCl_2$

4.2.2 MATERIALS AND EQUIPMENT

1. Disposable 15-ml polypropylene sterile tubes
2. Pasteur pipettes plugged with cotton
3. Forceps
4. Centrifuge

4.2.3 PROCEDURE

1. Place the worms in the earthworm farm, which basically consists of a covered styrofoam box with bedding up to the bottom of the ventilation holes, 2 weeks before the experiment. Then, sprinkle earthworm food on top of the bedding. The farm is kept at +10°C in the dark.

2. On the day of the experiment, rinse each worm with cold saline. Then, apply the following treatment: (1) Massage the posterior quarter length of the worm with forceps to expel the content from the lower gut and to minimize fecal contamination of the extrusion fluid. (2) Place each worm, one at a time, in a 15-ml tube containing 3 ml of extrusion medium. Incubate for 4 min at room temperature, keeping the tube on the side to avoid killing the worm.

3. Place the extruded cells (the 3 ml of extrusion fluid) in 15-ml tubes containing 11 ml of LBSS without calcium and keep on ice. Wash the cells three times in LBSS (without calcium). Each time, the cells are recovered following centrifugation at $250 \times g$ for 15 min at 4°C. Resuspend the cell pellet in 0.5 ml of LBSS with calcium.

4.3   HEMOCYTE COLLECTION IN THE MOLLUSK

4.3.1   REAGENT

1. Previously hemolymph will be collected

4.3.2   MATERIALS AND EQUIPMENT

1. 10-cc syringes
2. 22-gauge needles
3. Disposable 15-ml polypropylene sterile tubes
4. Laminar flow hood

4.3.3   PROCEDURE

1. Hemocytes are collected by puncturing the hemolymph from the posterior adductor muscle sinus.
2. Cells are transferred to tubes and washed. Cells are recovered following centrifugation at $250 \times g$ for 10 min.

4.4   CELL SUSPENSION FROM PERITONEAL EXUDATE

4.4.1   RAT

4.4.1.1   Reagents

1. HBSS containing 10 mM HEPES
2. RPMI 1640 with 10 U/ml heparin
3. 70% isopropanol
4. Fetal calf serum

4.4.1.2 Materials and Equipment

1. Sterile scissors
2. Sterile tissue forceps, hemostater forceps
3. 10-ml pipette
4. Disposable 15-ml (mouse) and 50-ml (rat) polypropylene sterile tubes
5. CO_2

4.4.1.3 Procedure

1. Euthanize the rat by CO_2 anoxia and dampen the fur with isopropanol.
2. Lay the rat on its back with its anterior abdomen exposed.
3. Lift the skin upward from the anterior abdominal wall with tissue forceps, and with scissors snip through the skin making an incision.
4. Strip back the fur and avoid touching the exposed abdominal wall.
5. Using a straight hemostatic forceps, lift and keep up the abdominal wall with the left hand, while with the right hand, using a 10-ml plastic pipette, inject 30.0 ml of RPMI 1640 with heparin into the peritoneal cavity, taking care to avoid puncturing the gut. If puncture occurs, the rat and the pipette should be discarded because of the bacterial contamination.
6. Circulate the injected fluid by gently massaging the abdomen on the sides of the rat and/or by aspirating the fluid in and out with the 10-ml plastic pipette.
7. Aspirate the fluid gently and transfer it to a 50-ml tube dipped in ice. Fill the tube with the RPMI 1640 with heparin.
8. Centrifuge the exudate at $250 \times g$ for 10 min at 4°C.
9. Resuspend the pellet in 1.0 ml of RPMI 1640 supplemented with 10% (v/v) fetal calf serum.

4.4.2 BIRDS

4.4.2.1 Reagents

1. HBSS containing 10 mM HEPES
2. RPMI 1640 with 10 U/ml heparin
3. Density gradient (1.077 g/ml)

4.4.2.2 Materials and Equipment

1. Sterile scissors
2. Sterile forceps
3. Syringes (50 ml)
4. Needles (18 gauge)
5. Polypropylene tubes (50 ml)
6. Nylon wool

4.4.2.3   Procedure

1. Harvest the cells by flushing the cavity with 30 ml of cold RPMI 1640 with heparin, being careful not to damage blood vessels or organs. Filter the peritoneal solution through nylon wool to remove fat and tissue particles; then centrifuge at $200 \times g$ for 10 min at 4°C.
2. Resuspend the cells in 3 ml of HBSS and layer on 3 ml of density gradient. Centrifuge at $100 \times g$ for 10 min (or longer if necessary) to remove any red blood cells (RBCs).
3. Collect the cells at the interface, dilute with HBSS, and centrifuge at $200 \times g$ for 6 min. Wash the cells a second time and resuspend in RPMI.

Note: To obtain a higher number of macrophages, thioglycolate can be injected intraperitoneally 2 days prior to collection of peritoneal macrophages.

4.5   ISOLATION OF WHITE BLOOD CELLS FROM PERIPHERAL BLOOD

4.5.1   MAMMALS

4.5.1.1   Reagents

1. HBSS containing 10 mM HEPES
2. RPMI 1640 with 10 U/ml heparin
3. NH_4Cl 0.15 M at 37°C; the temperature is crucial
4. RPMI 1640 with 10% fetal calf serum, 100 U/ml penicillin, and 100 µg/ml streptomycin

4.5.1.2   Materials and Equipment

1. Disposable 15-ml (mouse) and 50-ml (rat) polypropylene sterile tubes
2. Sterile Pasteur pipettes plugged with cotton

4.5.1.3   Procedure

White blood cells can be obtained from blood by lysing RBC or by centrifugation over a density gradient (d = 1.077 g/ml) where RBC and polymorphonuclear cells are centrifuged to the bottom of the tube. However, if both types of cells are required, lysis of RBC is preferred. To obtain polymorphonuclear cells, a gradient with a density of 1.113 g/ml can be used. Following centrifugation, two bands are obtained. The top band contains mononuclear cells, while the lower band contains polymorphonuclear cells.

A. Lysis of RBCs

1. In a 50-ml tube, add 3 ml of blood and 45 ml of 0.15 M NH$_4$Cl and incubate for 5 min at room temperature.*
2. Centrifuge 10 min at 275 × g at room temperature.
3. Discard the supernatant, add 10 ml of HBSS, transfer in a 15-ml tube, and wash twice with HBSS.

B. Centrifugation over a gradient

1. Dilute the blood 1:1 with HBSS.
2. Carefully layer the diluted blood over an equal volume of gradient with appropriate density (mononuclear cells: 1.077 g/ml; polymorphonuclear cells: 1.113 g/ml). Do not mix.
3. Centrifuge 30 to 40 min at 400 to 500 × g at room temperature.
4. Remove the supernatant and discard.
5. Harvest the cell layer and transfer to a 15-ml tube.
6. Add HBSS and centrifuge at approximately 300 × g for 10 min at room temperature.
7. Discard the supernatant and resuspend the cell pellet with RPMI supplemented with 10% fetal calf serum.

4.5.2 Birds

4.5.2.1 Reagents

1. HBSS
2. RPMI 1640 supplemented with 10% (v/v) fetal calf serum, 100 U/ml penicillin + 100 Ug/ml streptomycin
3. Heparin at 10 U/ml

4.5.2.2 Materials and Equipment

1. Syringes (10 ml)
2. Needles (18 gauge)
3. Disposable 15-ml or 50-ml polypropylene sterile tubes
4. Pasteur pipettes plugged with cotton

4.5.2.3 Procedure

1. Take blood through a heparinized syringe and transfer to a polypropylene tube. Then centrifuge the blood at low speed at 50 × g for 20 min at room temperature.

---

* An alternative method is given in Section 14.4.

2. Discard the erythrocyte cell pellet and collect the leukocyte-rich supernatant. Centrifuge the supernatant at $400 \times g$ for 10 min and resuspend it in HBSS.
3. Wash the leukocytes twice in HBSS and resuspend in supplemented RPMI 1640 for cultures.

4.5.3 FISH

4.5.3.1 Reagents

1. RPMI 1640 with 100 U/ml penicillin, 100 µg/ml streptomycin, and 10 mM HEPES
2. Gradient of 1.077 g/ml at room temperature to purify mononuclear cells (the temperature is very important; if too cold or warm, the gradient will change affecting its performance. The optimal temperature is 4°C)

4.5.3.2 Materials and Equipment

1. 10.0-ml pipettes
2. Disposable 15-ml or 50-ml polypropylene sterile tubes
3. Pasteur pipettes plugged with cotton

4.5.3.3 Procedure

1. Dilute the blood 1:10 with RPMI 1640.
2. Carefully layer the diluted blood over an equal volume of gradient with a density of 1.077 g/ml. Do not mix.
3. Centrifuge 15 to 20 min at $700 \times g$.
4. Harvest the cell layer and transfer to a 15-ml tube.
5. Wash the leukocytes once and resuspend in RPMI 1640.

4.6 CELL SUSPENSIONS FROM LYMPHOID ORGANS

The techniques described in this section can be applied to the spleen, thymus, lymph nodes, and head kidney.

4.6.1 REAGENTS

A. Mammals

1. HBSS containing 10 mM HEPES

B. Fish

1. HBSS supplemented with 100 U/ml penicillin, 100 µg/ml streptomycin, and 10 U/ml heparin
2. RPMI 1640 with 100 U/ml penicillin, 100 µg/ml streptomycin, and 10 mM HEPES

C. Rodents

1. RPMI 1640 with 100 U/ml penicillin, 100 µg/ml streptomycin, and 10 mM HEPES

4.6.2 MATERIALS AND EQUIPMENT

A. Mammals

1. Disposable 15-ml (mouse) and 50-ml (rat) polypropylene sterile tubes
2. Sterile Pasteur pipettes plugged with cotton
3. Sterile Pasteur pipettes plugged with nylon wool
4. Sterile tissue forceps
5. Sterile scissors
6. Sterile straight Hemostatic Forceps
7. Sterile plastic disposable pipettes
8. Sterile petri dish

B. Fish

1. Sterile cell strainer
2. Sterile Pasteur pipettes plugged with cotton
3. Disposable 15-ml polypropylene sterile tubes

C. Rodents

1. Petri dishes (47 mm diameter)
2. Cell strainer
3. Scissors
4. Forceps
5. Barrel of a 5- or 10-ml syringe
6. 15-ml polypropylene tubes
7. Pasteur pipettes plugged with cotton

4.6.3   PROCEDURE

A. Mammals

1. Remove the thymus, spleen, or lymph node as described in Chapter 3.
2. Mince the organ with two forceps by squeezing the spleen back and forth in a petri dish containing HBSS.
3. Add a volume of approximately 10 ml of HBSS to the culture dish and homogenize the cell suspension with a sterile Pasteur pipette plugged with cotton wool.
4. Then pass the cell suspension through a sterile Pasteur pipette plugged with nylon wool and recover it in a disposable 15-ml or 50-ml polystyrene tube.
5. Adjust the volume to approximately 15 ml or 50 ml with HBSS.

B. Fish

1. Remove the thymus, spleen, or head kidney as described in Chapter 3.
2. Using a sterile cell strainer on a petri dish, gently crush the organ, rinsing with HBSS (10 U/ml heparin) to wash the cells through the mesh.
3. Transfer the suspension in a 15-ml tube, add 12 ml medium, and centrifuge 10 min at $300 \times g$.
4. Discard the supernatant and resuspend the cell pellet in 3 ml of RPMI.

C. Rodents

1. Collect organ into 2 ml (thymus) or 5 ml (spleen) of RPMI 1640 and keep on ice (see Chapter 3).
2. Add a small volume of medium, approximately 2 ml, to the plastic dish. Transfer tissue into the dish. Using scissors, cut tissue into three to four pieces. Using the cell strainer, gently crush the organ until all pieces are disintegrated into single cells. *Note:* to minimize cell destruction you must avoid grinding while pressing with the barrel.
3. With a disposable Pasteur pipette, collect all cells and transfer to a 15-ml polypropylene tube. Allow it to stand on ice for 5 min to sediment connective tissue. Transfer top layer of cells into a clean tube and centrifuge for 10 min at $150 \times g$. Discard supernatant and resuspend cells to 3 ml (thymus) or 5 ml (spleen) medium.
4. Keep cells at room temperature until needed.

SUGGESTED READING

DeSwart, R.L., Kluten, R.M.G., Huizing, C.J., Vedder, L.J., Reijnders, P.J.H., Visser, I.K.G., Uyt de Haag, F.G.C.N., and Osterhaus, A.D.M.E., 1993. Mitogen and antigen induced B and T cell responses of peripheral blood mononuclear cells from the harbour seal (*Phoca vitulina*), *Vet. Immunol. Immunopathol.*, 37, 217–230.

De Koning, Y. and Kaattari, S., 1992. An improved salmonid lymphocyte culture medium incorporating plasma for *in vitro* antibody production and mitogenesis, *Fish Shellfish Immunol.*, 2, 275–285.

Fugère, N., Brousseau, P., Krzystyniak, K., Coderre, D., and Fournier, M., 1996. Heavy metal specific inhibition of phagocytosis and different *in vitro* sensitivity of heterogeneous coelomocytes from *Lumbricus terrestris* (Oligochaeta), *Toxicology,* 109, 157–166.

Hunt, S.V., Preparation of lymphocytes and accessory cells, in *Lymphocytes: A Practical Approach,* G.G.B. Klaus, Ed., IRL Press, 1987, 1–34.

5 Assessment of Cell Viability

5.1 INTRODUCTION

The preparation of a cell suspension whether from blood, body fluids, or organs is stressful for the cells. Consequently, some of the cells may die. Furthermore, when working directly from a cell culture, it is important to know the state of health of the cells prior to use. Two dyes are commonly used to evaluate cell viability: trypan blue dye and ethidium bromide with acridine orange. Viability can also be assessed by flow cytometry with propidium iodide (PI).

5.2 DETERMINATION OF CELL VIABILITY AND CONCENTRATION BY TRYPAN BLUE DYE EXCLUSION

5.2.1 REAGENTS

1. Hanks' balanced salt solution (HBSS) containing 10 mM of HEPES
2. Trypan blue dye solution at 0.4% (w/v) in phosphate buffer saline (PBS)
3. 70% ethanol
4. Distilled water

5.2.2 MATERIALS AND EQUIPMENT

1. Hemacytometer (Neubauer) and thick coverslip
2. Manual counter with two different units
3. Pasteur pipettes
4. Bulbs
5. Optic microscope
6. Disposable 12 × 75 mm tubes

5.2.3 PROCEDURE

1. Clean the hemacytometer and the coverslip with ethanol and rinse with distilled water. Dry completely.
2. Lay the coverslip across the two counting areas and press on the sides.
3. In a disposable tube, add 0.8 ml of the buffer plus 0.1 ml of cell suspension plus 0.1 ml of the trypan blue solution.* Gently mix with a pipette.

---

* The dilution with the dye could be changed according to the expected cell concentration.

4. Introduce the cell suspension onto the counting area under the coverslip using the V notch. Areas will be filled by capillary action. A volume of approximately 0.01 ml is needed.

5. Counting must take place within 5 to 15 min after step 3. Observe the preparation under the microscope (40×). Lymphocytes are larger than erythrocytes and platelets. Dead cells are blue and usually larger than live cells.

6. The hemacytometer consists of 25 large squares enclosed by triple lines, and each of them is divided into 16 smaller squares.

7. Count the number of viable and dead cells in five large squares using the counter to record numbers. If the total is less than 100 cells, count the number of cells in the total counting area (25 large squares). Count the two counting areas of the hemacytometer for the same sample and calculate a mean value. A maximum difference of 10% between the two counts is acceptable.

8. Calculate the concentration of the cell suspension using the following equation:

$$[\ ]\ \text{cells/ml} = X \times 10^4 \times Y$$

where

X = total number of cells in 25 squares
10^4 = transform the volume of 0.1 mm³ into 1 cm³
Y = dilution factor

9. Calculate viability using the following equation:

$$\%\ \text{viability} = A/B \times 100$$

where

A = mean value of the number of viable lymphocytes
B = total number of viable and dead lymphocytes

Viability of greater than 80% is generally acceptable.

5.3 DETERMINATION OF CELL VIABILITY AND CELL CONCENTRATION WITH ETHIDIUM BROMIDE AND ACRIDINE ORANGE

5.3.1 REAGENTS

1. HBSS containing 10 mM of HEPES
2. Stock solution of ethidium bromide/acridine orange (100×).
3. 50 mg ethidium bromide

4. 15 mg acridine orange
5. Working solution of ethidium bromide/acridine orange

Dissolve both reagents in 1.0 ml of 95% ethanol. Add 49 ml distilled water. Mix well, divide into 1.0-ml aliquots, and freeze at –20°C. This stock can be kept frozen for at least 3 months. Prepare the working solution of ethidium bromide/acridine orange by thawing a 1.0-ml aliquot of the 100× stock solution diluted 1/100 in HBSS or other isotonic medium. Mix well and store in an amber bottle at 4°C for up to 1 month.

5.3.2 MATERIALS AND EQUIPMENT

1. Fluorescence microscope
2. Hemacytometer chamber and coverslip
3. Pasteur pipettes
4. Bulbs
5. Disposable 12 × 75 mm tubes

5.3.3 PROCEDURE

1. Adjust an aliquot of the cell suspension to an estimated 1 to 5×10^6 cells/ml in HBSS or other isotonic medium.
2. Add equal volumes (25 µl) of cell suspension and ethidium bromide/acridine orange solution in a tube and mix gently.
3. Place a small aliquot under the coverslip on a hemacytometer slide.
4. Observe cells initially under the microscope using visible light (100 to 400×). Adjust the diaphragm to reduce light and still keep the hemacytometer grid visible.
5. Keeping the visible light on, switch to fluorescence and observe the cells. Live cells exhibit green fluorescence (with acridine orange) and dead cells exhibit orange fluorescence (with ethidium bromide).
6. A minimum of 100 cells must be counted. Take the count of two counting areas for the same sample and calculate a mean value. A maximum difference of 10% between the two counts is acceptable.
7. Calculate the concentration by using the following equation:

$$[\ ]\text{cells/ml} = X \times 10^4 \times Y$$

where

X = number of cells in 25 squares
Y = dilution factor
10^4 = to transform the volume of 0.1 mm³ into 1 cm³

8. Calculate viability by using the following equation:

$$\% \text{ viability} = A/B \times 100$$

where

A = mean value of the number of viable lymphocytes
B = summation of viable and dead lymphocytes

A percentage of viability greater than 80% is generally acceptable.

5.4   FLOW CYTOMETRIC ASSESSMENT OF CELL VIABILITY

5.4.1   REAGENTS

1. Dulbecco's PBS
2. PI* stock solution of 1 mg/ml deionized water in a 15-ml tube (solution can be filtered after dissolution of PI in water)
3. A 10× working solution of PI (10 µg/ml in PBS) prepared in a 15-ml tube

5.4.2   MATERIALS AND EQUIPMENT

1. Disposable 5-ml polystyrene (12 × 75 mm) test tubes
2. Disposable 15-ml polypropylene tubes
3. Flow cytometer

5.4.3   PROCEDURE

1. Adjust an aliquot of the cell suspension to an estimated 1 to 5×10^6 cells/ml in PBS or other isotonic medium.
2. Add 900 µl of cell suspension in a 12 × 75 mm tube; then add 100 µl of PI 10× working solution — or 10% PI 10× (v/v) to the cell suspension volume — and mix gently.
3. Analyze with the flow cytometer within 1 h.
4. Assay the viability using intracellular red fluorescence (FL2). A logarithmic amplification of red fluorescence (FL2) signals is used. Event triggering is based on forward and side light scatter. Viable cells are identified as those with low red fluorescence and dead cells as those with high red fluorescence on the FL2 parameter.

---

* PI is a possible carcinogen and should be handled with caution.

5.5 WORKING SHEET: CELL VIABILITY AND CONCENTRATION BY TRYPAN BLUE DYE EXCLUSION — OR ETHIDIUM BROMIDE/ACRIDINE ORANGE

Method used:	Trypan Blue: ☐ E.B. / A. O. ☐	Study title: _____ Prepared by: _____	Date: _____

Animal #:	CELL COUNT								# of squares counted	Susp. Volume (ml)	Dilution	% Viability	[cell] (cells/ml)	Total # of cells
	Live				Dead									
	1	2	Mean	S.D.	1	2	Mean	S.D.						

Page ___ of ___

SUGGESTED READING

Parks, D.R., Bryan, V.M., Oi, V.T., and Herzenberg, L.A., 1979. Antigen-specific identification and cloning of hybridomas with a fluorescence-activated cell sorter, *Proc. Natl. Acad. Sci. U.S.A.*, 76, 1962–1966.

6 Cell Cryopreservation

6.1 INTRODUCTION

This chapter will describe how to freeze cell cultures in liquid nitrogen and how to thaw them for further use.

6.2 REAGENTS

1. RPMI 1640 with 25 mM HEPES, L-glutamine, 100 U/ml penicillin, 100 µg/ml streptomycin, and 15% heat-inactivated fetal bovine serum (FBS)
2. Dimethyl sulfoxide (DMSO)
3. Liquid nitrogen

6.3 MATERIALS AND EQUIPMENT

1. Cryovials of 1.2, 2, or 5 ml
2. Aluminium canes
3. Freezing tray
4. Freezing container according to volume of vials
5. Disposable 15-ml polypropylene tubes
6. Tissue culture flask
7. −80°C freezer
8. Liquid nitrogen container
9. Cryobox
10. 37°C water bath
11. Incubator set at 37°C, 5% CO_2, in humidifed atmosphere
12. Materials and reagents to determine cell viability (see Chapter 5)

Note: A mixture of DMSO and FCS produces an exothermic reaction (~50°C); FCS and RPMI should be refrigerated to 4°C before use.

6.4 PROCEDURE

6.4.1 FREEZING OF CELL CULTURES

1. Grow cells to late log phase and microscopically examine cultures for healthy growth and absence of bacterial contamination.
2. Pool 40.0 ml of suspension from culture flasks into 50.0-ml Corning centrifuge tubes. Mix well to obtain a uniform suspension. Remove 0.5 ml for counting and viability determination. Discard if viability is <90%.
3. Perform cell counts and viability determinations.
4. Centrifuge to pellet cells at $300 \times g$ for 10 min.
5. Resuspend cell pellet in freezing medium (10% DMSO in complete medium) to approximately 2.5×10^6 cells/ml.
6. Dispense 1.0-ml aliquot of cell suspension into prelabeled cryotubes using an eppendorf repeater pipette. Cap tubes tightly.
7. Transfer cryotubes onto the freezing tray of the liquid nitrogen storage tank, starting at the most distant location in the storage tank of liquid N_2.
8. Allow tray to stand in the liquid N_2 storage tank for 30 min.
9. Keep lowering the tray step by step until final position is reached, allowing it to stand for 30 min at each position.
10. Transfer cryotubes into prelabeled and prechilled aluminum canes.
11. Store canes in the liquid phase of liquid nitrogen storage tank. Record the position of canes in the logbook provided for each storage tank.
12. Tests for mycoplasma contamination must be performed on cell batches.

6.4.2 THAWING OF CELLS

1. Thaw cell from frozen seed stock as follows: Place frozen ampoule of vial directly into a 37°C water bath and agitate vigorously. As soon as specimen is melted, remove the ampoule/vial from the water bath and swab with 70% ethanol. All operations from this point on are to be carried out under the laminar flow hood.
2. Break the neck of the ampoule between several folds of sterile gauze.
3. Transfer the contents of the ampoule/vial into a prelabeled 15.0-ml centrifuge tube.
4. Add 10.0 ml of complete culture medium slowly to the cell suspension — 10.0 ml over about 2 min added dropwise at the start and then a little faster is ideal for gradual dilution of the cells. This gradual dilution is very important, as sudden dilution of DMSO contained in the cryopreservative can cause severe osmotic damage and reduce cell survival.
5. Centrifuge the diluted suspension at approximately $300 \times g$ for 10 min.
6. Discard supernatant and resuspend cell pellet in 1.0 ml of culture medium. Remove 0.5 ml for counting and viability determination.
7. Adjust cell suspension to 1×10^5 viable cells/ml.

8. Dispense into appropriately sized tissue culture flasks:
   For 25 cm^2 — NOT to exceed 12.0 ml
   For 75 cm^2 — NOT to exceed 50.0 ml
9. Place flasks in 37°C, 5% CO_2 incubator, lying flat as in monolayer cell culture. Loosen the cap of each flask by turning it counterclockwise ¾ of a revolution.
10. On Day 3 or 4 (late log phase of growth cycle for K562) subculture by splitting cells 1:10 with fresh culture medium.

SUGGESTED READING

De Boer, M., Reijneke, R., Van De Griend, R.J., Loos, J.A., and Roos, D., 1981. Large-scale purification and cryopreservation of human monocytes, *J. Immunol. Methods,* 43, 225–239.

Fujiwara, S., Akiyama, M., Yamakido, M., Seyama, T., Kobuke, K., Hakoda, M., Kyoizumi, S., and Jones, S.L., 1986. Cryopreservation of human lymphocytes for assessment of lymphocyte subsets and natural killer cytotoxicity, *J. Immunol. Methods,* 90, 265–273.

Golub, S.H., Sulit, H.L., and Morton, D.L., 1975. The use of viable frozen lymphocytes for studies in tumor immunology, *Transplantation,* 19(3), 195–202.

Hansen, J.-B., Halvorsen, D.S., Haldorsen, B.C., Olsen, R.S., Jursen, H., and Kierulf, P., 1995. Retention of phagocytic functions in cryopreserved human monocytes, *J. Leukocyte Biol.,* 57, 235–241.

Ludgatem, M.E., Dryden, P.R., Weetman, A.P., and McGregor, A.M., 1983. T-cell subsets analysis of cryopreserved human peripheral blood mononuclear cells, *Immunol. Lett.,* 7, 119–122.

Prince, H.E. and Lee, C.D., 1986. Cryopreservation and short-term storage of human lymphocytes for surface marker analysis. Comparison of three methods, *J. Immunol. Methods,* 93, 15–18.

Tollerud, D.J., Brown, L.M., Clark, J.W., Neuland, C.Y., Mann, D.L., Pankiw-trost, L.K., and Blattner, W.A., 1991. Cryopreservation and long-term liquid nitrogen storage of peripheral blood mononuclear cells for flow cytometry analysis: effects on cell subset proportions and fluorescence intensity, *J. Clin. Lab. Anal.,* 5, 255–261.

Van der Meulen, F.W., Reis, M., Stricker, E.A., Van Elven, E., and Von Dem, B.A.E., 1981. Cryopreservation of human monocytes, *Cryobiology,* 18, 337–343.

7 Phagocytosis Functional Assay

7.1 INTRODUCTION

The first line of cellular defense in invertebrates and vertebrates is phagocytosis of foreign substances. Phagocytosis is performed by phagocytes; the nomenclature varies with their location in the body (e.g., blood neutrophils and tissue macrophages) and with the species (e.g., earthworms, coelomocytes, and birds, heterophils). Nevertheless, they all have the same function: recognition, engulfment, and destruction of material recognized as foreign.

In this chapter, we describe phagocytosis using flow cytometry and murine peritoneal macrophages as a model (consult Section 7.7 for species-related differences in the protocol). In this assay, cells are exposed to fluorescent beads or yeast, and, after an appropriate period of incubation (the cell function and the species must be considered), the number of engulfed beads is determined by flow cytometry or visually on a microscope. The use of a gradient prior to the flow cytometric evaluation helps to remove free beads or beads slightly adhered to the surface of the cells. The different control groups run during the assay in flow cytometry ensure that the active phenomenon of phagocytosis is evaluated. This assay may be run in parallel with the oxidative burst determination and/or the antigen-processing assay, allowing for a more complete evaluation of the role of macrophages in natural and specific immune responses. In addition, this assay can follow the phagocytic function following *in vivo* treatment of laboratory animals with chemicals.

The phagocytic function using flow cytometry for murine peritoneal macrophages is performed with the following suggested material and equipment (consult Section 7.1 for species-related differences in protocol). Typical FACScan results are presented in Figures 7.1 and 7.2.

7.2 REAGENTS

Reagents necessary to collect and prepare cell suspensions (see Chapters 1 through 5):

1. Hanks' buffered saline solution (HBSS) supplemented with 25 mM HEPES
2. RPMI 1640 containing 25 mM HEPES buffer and L-glutamine, phenol red, and sodium bicarbonate
3. RPMI 1640 supplemented with 10% fetal calf serum (FCS) (v/v), 100 U penicillin, 100 µg streptomycin/ml

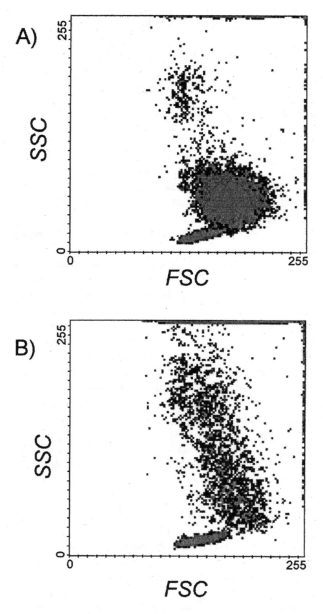

FIGURE 7.1   Scattergram of rat peritoneal leukocytes. A scattergram of the flow cytometric profile of rat peritoneal leukocytes (A). Cells are displayed according to their size (FSC; x axis) and complexity (SSC; y axis). The peritoneal exudate contained some lymphocytes which were seen as small cells with low complexity, while macrophages are clearly more complex or granular. In the analysis, we must keep in mind that activation of cells in phagocytosis will result in an increase of granularity and the subsequent modification of FSC and SSC properties (B).

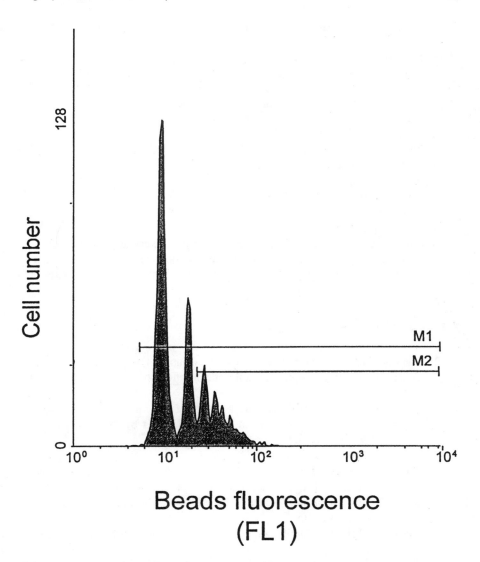

FIGURE 7.2 Histogram of phagocytosis. Typical histogram of fluorescence emitted by beads engulfed by the phagocytes. The marker M1 represents the percentage of phagocytic cells with one bead and more. The marker M2 represents the percentage of phagocytic cells with three beads and more.

4. RPMI 1640 supplemented with 10% FCS (v/v), 100 U penicillin, 100 μg streptomycin/ml, and 0.2% (w/v) sodium azide (NaN_3)
5. RPMI 1640 + 3% bovine serum albumin (BSA) (w/v) at 4°C: centrifugation gradient
6. Sheath fluid
7. 0.5% formalin in sheath fluid

7.3   MATERIALS AND EQUIPMENT

Materials and equipment necessary to collect and prepare cell suspensions (see Chapters 1 through 5):

1. 12 × 75 mm polystyrene round-bottom tubes
2. Disposable 15-ml polypropylene tubes
3. Centrifuge set at 4°C
4. Incubator set up at 37°C, 5% CO_2 in a humidified atmosphere
5. Rotating agitator
6. Ice bath with lid
7. Automatic pipettes and tips
8. Ultrasonic bath
9. Pasteur pipettes and bulbs
10. Fluorescent carboxylate latex beads — 1.5 μm (excitation 488 nm, emission 520 nm) for assays with peritoneal macrophages (the following procedures) and latex beads: 1.0 μm, latex for assays with neutrophils or monocytes (blood)

7.4   PROCEDURE (see Figure 7.3)

1. Harvest the peritoneal macrophages (see Chapter 4).
2. After the last wash, adjust the cell concentration to 1.0×10^6 cells/ml in HBSS.
3. For each cell suspension, prepare aliquots of 2.0 ml each in disposable 15-ml polypropylene tubes. Then recover the cells by centrifugation at $250 \times g$ for 10 min.
4. Resuspend the cells as follows:

   Tube A: 2.0 ml of supplemented RPMI 1640, without NaN_3: experimental and/or positive control cells (untreated cells);

   Tube B: 2.0 ml of supplemented RPMI 1640, without NaN_3: negative control cells;

   Tube C: 2.0 ml of supplemented RPMI 1640, with NaN_3: negative control cells.

   Incubate tubes A and C at 37°C for 30 min before the addition of the beads. Incubate tube B in wet ice for 30 min before the addition of the beads.
5. After this first incubation, add the beads to the tubes and incubate at 37°C or in wet ice (following the same code as in Step 4) under mild agitation (the agitator should be put in the incubator).

Note: Use the following equation to calculate the concentration of latex beads per milliliter. The number of beads per milliliter must be determined for each new bottle using the following equation. In addition, before adding the beads into the phagocyte suspension, sonicate them for 5 min at room temperature

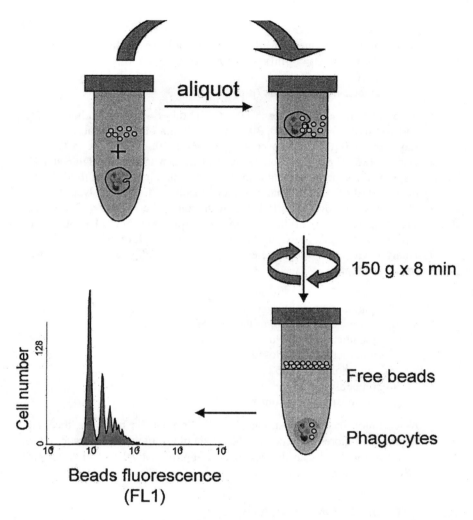

aliquot

150 g x 8 min

Free beads

Phagocytes

Beads fluorescence
(FL1)

FIGURE 7.3 Principle of the phagocytosis. Cells are incubated with microspheres made of latex-containing fluorescein (FITC). Following an appropriate incubation period, the suspension comprising cells and microspheres is layered over a density gradient (3% BSA). By centrifugation, the cells are pelleted, whereas nonphagocytosed beads stay at the surface of the gradient. Cells are then analyzed by flow cytometry where the intensity of the fluorescence is related to the number of microspheres phagocytosed by phagocytes.

in the ultrasonic bath to get rid of doublet and triplet beads. The final proportion of latex beads per macrophage should be 100:1.

$$\text{\# beads} / \text{ml} = \frac{6W \times 1012}{\pi \times \partial \times \beta^3}$$

where

W = grams of polymer per ml (0.025 g for a latex solution of 2.5%)
∂ = polymer density (g/ml): 1.05 for latex
π = 3.1416
β = diameter of latex beads (μm)

6. Take a 500-μl sample of each tube (37°C tubes with and without NaN$_3$ and wet ice control) after 0, 30, 60, and 90 min of incubation.
7. Layer the 500-μl sample over 4 ml RPMI 1640 + 3% BSA (4°C) in a disposable 5-ml polystyrene tube and centrifuge at $150 \times g$ for 8 min at 4°C.
8. Aspirate the supernatant and resuspend cell pellet in 500 μl of fluid sheath or in 500 μl of 0.5% formalin in fluid sheath for further analysis.
9. Cap tubes and cover with foil to protect from light; keep at 4°C until the sample is acquired with flow cytometer. Acquisition should be completed within 24 h if cells are resuspended in 0.5% formalin in FACS Flow.
10. Analyze with the flow cytometer. The following configuration is suggested for a FACScan:

FSC parameter	Linear mode (1.30)
SSC parameter	Logarithmic mode/detector at 177
FL1 parameter	Logarithmic mode/detector at 382
FL2 parameter	Closed
FL3 parameter	Closed
FSC threshold	152
No compensation	
(Debris must be gated out.)	

Note: Fluorescence adjustment with beads must be made on the FL1 parameter. Fluorescence peak of beads should fit in the first decade of the FL1 scale with a mean fluorescence of approximatively 10. (Do not put FSC threshold during calibration.)

7.5   SPECIAL RECOMMENDATIONS

1. *Negative control groups:* A 4°C control without sodium azide and a 37°C control with sodium azide should be made for all treated/untreated cell suspensions. Ensure that control cells never phagocytize, but remain alive for the duration of the experiment in order to measure adherence of beads on the cell surface. Viability of control cells should be determined at the end of the experiment.
2. *Neutrophils, monocytes, and macrophages:* A high concentration of neutrophils in cell suspension will substantially increase the phagocytic response due to biological properties of this cell population. In addition, dead macrophages, neutrophils, or monocytes will slowly degenerate into small fragments that can erroneously be gated as live cells. Therefore, the studied cell population must be defined and gated appropriately (Figure 7.1).
3. *Lymphocytes:* Must be gated out.

7.6 ANALYSIS OF RESULTS

At least three types of analysis can be performed with this technique. Each type of analysis should be made only with gated phagocytic cell populations, peritoneal macrophages, for example (Figure 7.1). Background negative control fluorescence usually does not exceed the first decade of fluorescence (one bead peak). Results can be illustrated relatively from the time of incubation from the beginning of the process (i.e., from 0 min to 90 min or more). The phagocytic activity peaks at 60 or 90 min and tends to drop afterward.

Results can be recorded three ways: first, the percentage of phagocytic cells with one bead or more is calculated (Figure 7.2). With an FL1 histogram on a gated population, separate cells of the first decade and more (fluorescence $\leq 10^1$) from the inactive cells (fluorescence $\geq 10^1$). Then, compare the percentage of gated events for one bead or more between control and experimental tubes.

Second, the percentage of phagocytic cells with three beads or more is calculated (Figure 7.2). With an FL1 histogram on a gated population, select the third peak, which tends to merge with following peaks. Then, compare the percentage of gated events for three beads or more between control and experimental tubes.

Third, the mean number of beads per gated cell is calculated. It assumes that each multiple of the first decade of fluorescence (one bead) reflects the addition of the same multiple of beads. For example, 20 units of fluorescence should represent, in fact, two beads engulfed in a cell, since one bead represents 10 units of fluorescence. Since the population of phagocytic cells is already gated (live gate or analysis gate), the percentage of cells actually involved in phagocytosis is reflected by the percentage of cells that engulfed one bead or more (% gated events; Figure 7.2 — statistical analysis M2). Therefore, the mean number of beads per gated cell is found by multiplying the percentage of gated cells by the mean fluorescence units of the cells that engulfed one bead and more, and by dividing the total by the fluorescence unit of the first peak representing one bead. Then, the same procedure is performed with the negative control sample. Since the negative control sample represents adherence of beads on surface membrane, the mean number of beads per cell from the negative control sample should be substracted from the experimental mean number of beads per cell.

Experimental   –   Negative Control

$$\text{Mean number} = \frac{\%\text{gated events} \times \text{mean fluo. of cells} (\geq 1 \text{ bead})}{\text{fluorescence unit of the first peak}} - \frac{\%\text{gated events} \times \text{mean fluo. of cells}}{\text{fluorescence unit of the first peak}} \quad \frac{(\geq 1 \text{ bead})}{\text{of beads per gated cell}}$$

Comments: The method may not differentiate between beads that are attached to the macrophage surface and those ingested inside the macrophage.

7.7 SPECIES-RELATED CHANGES IN PROTOCOL

Species	Reagents	Materials	Methods
Bovine	—	1. Use 1.0 μm latex beads.	1. Use blood neutrophils and/or monocytes; prepare samples according to Section 2.4. 2. Adjust cell concentration to 2.0 × 10^6 cells/ml after the last wash (step 2).
Duck	—	1. Use 1.0 μm latex beads. 2. Cells must be kept at 41°C.	1. Use blood neutrophils and/or monocytes. Prepare samples according to Section 2.4. 2. Monitor phagocytosis up to 90 min of incubation (0, 30, 60, and 90 min incubation periods).
Earthworm (*Lumbricus terrestris*)	—	1. Use 1.5 μm latex beads.	1. Use coelomocytes; prepare samples according to Chapter 4 (extrusion of coelomocytes from earthworm). 2. Monitor phagocytosis up to 18 h of incubation at 15°C.
Human	—	1. Use 1.0 μm latex beads.	1. Use blood neutrophils and/or monocytes; prepare samples according to Section 2.4. 2. Monitor phagocytosis up to 90 min of incubation (0, 30, 60, and 90 min incubation periods).
Mink	—	1. Use 1.0 μm latex beads.	1. Use blood neutrophils and/or monocytes; prepare samples according to Section 3.4. 2. Monitor phagocytosis up to 120 min of incubation (0, 30, 60, 90, and 120 min incubation periods).
Monkey	—		1. Use blood neutrophils and/or monocytes; prepare samples according to Section 2.4. 2. Monitor phagocytosis up to 90 min of incubation (0, 30, 60, and 90 min incubation periods).
Mussel	Hemolymph	1. Use 1.5 μm latex beads.	1. Use hemocytes; 18 h incubation periods.
Pig	—	1. Use 1.0 μm latex beads.	1. Use blood neutrophils and/or monocytes; prepare samples according to Section 3.4. 2. Monitor phagocytosis up to 60 min of incubation (0, 5, 10, 20, 30, 40, and 60 min incubation periods).
Rat	—	1. Use 1.7 μm latex beads.	1. Use blood neutrophils and/or monocytes or peritoneal macrophages; prepare samples according to Section 2.4. 2. Monitor phagocytosis up to 60 min of incubation (0, 5, 10, 20, 30, 40, and 60 min incubation periods).
Seal	—	1. Use 1.0 μm latex beads.	1. Use blood neutrophils. 2. Monitor phagocytosis up to 90 min of incubation (0, 30, 60, and 90 min incubation periods).
Trout (*Onchorhynchus mykiss*)	1. Use RPMI 1640 + 10% FBS + 3% BSA for centrifugation gradient	1. Use 2.0 μm latex beads.	1. Incubate for 18 h at 20°C; use head kidney phagocytes; prepare samples according to Section 3.4

7.8 TECHNIQUE-RELATED CHANGES IN THE PROTOCOL: PHAGOCYTOSIS WITH YEAST

7.8.1 REAGENTS

1. Mouse serum
2. Sabouraud's dextrose agar
3. RPMI with 10% mouse serum and 100 U/ml penicillin, 100 μg/ml streptomycin
4. Phosphate Buffer Saline (PBS)
5. HBSS
6. Formalin 10% in PBS

7.8.2 MATERIALS AND EQUIPMENT

1. 35-mm petri dishes
2. Sterile 11 × 22 mm glass coverslips
3. *Saccharomyces cervisiae*
4. Light microscope

7.8.3 PROCEDURE

A. *Culture of macrophages on coverslips*

1. Plate 5×10^5 or 1×10^6 cells in 1 ml supplemented RPMI onto 35 mm Falcon petri dish containing two 11 × 22 mm glass coverslips.
2. Gently shake the plate by hand to distribute the cells on the coverslips evenly.
3. Incubate for 2 h at 37°C in a 5% CO_2 incubator evenly.
4. Wash the plates and the coverslips three times with warm (37°C) RPMI to remove the nonadherent cells.
5. Place the coverslips back into a 35-mm dish.
6. Add 1 ml supplemented RPMI and keep in a 5% CO_2 incubator while preparing the yeast cells.

B. *Preparation of yeast cells*

1. Suspend 3-day-old cultures of cake yeast on Sabouraud's dextrose agar in PBS.
2. Wash the yeast cells three times in HBSS at $150 \times g$ for 15 min.
3. Count the yeast cells.
4. Adjust the yeast cells to 1×10^7 cells/ml.
5. Spin 10 ml of the yeast cell suspension (1×10^8 cells) at $150 \times g$ for 15 min.
6. Suspend the yeast cell pellet in 10 ml of RPMI containing 10% fresh mouse serum.
7. Incubate at 37°C for 30 min to allow for opsonization of the yeast cells.

8. Wash the yeast cells in HBSS three times and centrifuge at $150 \times g$ for 10 min.
9. Resuspend yeast cells in supplemented RPMI at 5×10^6 cells/ml or 1×10^7 cells/ml.

C. Phagocytosis assay

1. Aspirate off the RPMI media from the petri dishes containing the coverslips.
2. Add 1 ml of yeast cell suspension to the petri dish and gently shake to distribute the yeast cells evenly (to give a ratio of 1:10 or more of macrophages to yeast cells).
3. For time zero phagocytosis — immediately rinse the coverslips and dish with cold PBS three times to wash off unphagocytosed yeast cells.
4. Fix the coverslips by adding 1 ml of phosphate buffered 10% formalin to the petri dish.
5. Repeat Steps 3 and 4 after incubating the macrophages with the yeast cells for 30 and 60 min. Agitate the dishes at regular intervals.
6. Stain the fixed coverslips with eosin and hematoxylin (Histology Laboratory procedure) to identify the macrophages and the ingested yeast cells.
7. Mount the coverslips on microscope slides for light microscopy.

D. Data collection

1. Under a light microscope at 40× objective lens, count five fields per slide to determine the total number of macrophages/field and the number of macrophages that have ingested one or more yeast cells.
2. Each field to be counted should contain at least 70 to 100 macrophages.

7.9   STATISTICAL TESTS

Results can be analyzed using analysis of variance (ANOVA) statistical procedure, either with Tukey's or Scheffé's means differentiation tests. Include 37 and 4°C untreated/treated phagocytose results in matrix, in order to separate positive and negative control results from experimental results.

7.10 WORKING SHEET: PHAGOCYTIC FUNCTION

Animal #: _____ Experiment #: _____
Type: _____ Date: _____
Prepared by: _____

Harvested macrophages:

% Cell viability: _____ Number of cells harvested: _____

Cell suspension volume: _____ ml

Exact cell concentration: _____ cells/ml

FACS acquisition:

Data File #:

		Time (minutes)		
		0	30	60
Tube A)	(37°C)			
Tube B)	(4°C control)			
Tube C)	(37°C control *)			

Results:

% of phagocytic cells with 1 bead and more

		Time (minutes)		
		0	30	60
Tube A)	(37°C)			
Tube B)	(4°C control)			
Tube C)	(37°C control *)			

% of phagocytic cells with 3 beads and more

		Time (minutes)		
		0	30	60
Tube A)	(37°C)			
Tube B)	(4°C control)			
Tube C)	(37°C control *)			

Mean fluorescence x number of cells

		Time (minutes)		
		0	30	60
Tube A)	(37°C)			
Tube B)	(4°C control)			
Tube C)	(37°C control *)			

* with sodium azide

SUGGESTED READING

Fredrickson, A.G., Hatzis, C., and Srienc, F., 1992. A statistical analysis of flow cytometric determination of phagocytosis rates, *Cytometry,* 13, 423–431.

Gadebush, H.H., 1979. *Phagocytes and Cellular Immunity,* CRC Press, Boca Raton, FL, 164 pp.

Grand-Perret, T., Bravo-Cuellar, A., Orbach-Arbouys, S., Caraux, J., and Barot-Ciorbaru, R., 1990. Changes in murine peritoneal cell activity after administration of diethyldithiocarbamate, *Int. J. Immunother.,* 6(4), 215–220.

Métezeau, P., Ronot, X., Le Noan-Merdrignac, G., and Ratinaud, M.H., Eds., 1982. *La Cytométrie en Flux,* Vol. 1, MEDSI/McGraw-Hill, Ontario, Canada, 372 pp.

Molecular Probes, *How to Work with Latex Microspheres,* Molecular Probes, Inc., 1991, 11 pp.

Perticari, S., Presani, G., Mangiarotti, M., and Banfi, E., 1991. Simultaneous flow cytometric method to measure phagocytosis and oxidative products by neutrophils, *Cytometry,* 12, 687–693.

Perticari, S., Presani, G., and Banfi, E., 1993. A new flow cytometry assay for the evaluation of phagocytosis and the oxidative burst in whole blood, *J. Immunol. Methods,* 170, 117–124.

Pratten, M.K. and Lloyd, J.B., 1984. Phagocytic uptake of latex beads by rat peritoneal macrophages: comparison of morphological and radiochemical assay methods, *Biosci. Rep.,* 4, 497–504.

Rasmussen, S., 1991. *An Introduction to Statistics with Data Analysis,* W.H. Freeman and Company, Cole, CA, 707 pp.

Seymour, L., Schacht, E., and Duncan, R., 1991. The effect of size of polystyrene particles on their retention within the rat peritonal compartment, and on their interaction with rat peritoneal macrophages *in vitro, Cell Biol. Int. Rep.,* 15, 277–286.

Stewart, C., Lehnert, B., and Steinkamp, J.A., 1986. *In vitro* and *in vivo* measurement of phagocytosis by flow cytometry, *Methods Enzymol.,* 132, 183–191.

Stuart, A.E., Habeshaw, J.A., and Davidson, E., 1978. Phagocytes *in vitro*, in Weir, D.M., Ed., Handbook of Experimental Immunology: Cellular Immunology, Vol. 2, Elsevier, London, 31.1–31.30.

Thuvander, A., Norrgren, L., and Fossum, C., 1987. Phagocytic cells in blood from rainbow trout, *Salmo gairdneri* (Richardson), characterized by flow cytometry and electron microscopy, *J. Fish. Biol.,* 31, 197–208.

Van Furth, R., Van Zwet, L., and Leijh, P.C.J., 1978. *In vitro* determination of phagocytosis and intracellular killing by polymorphonuclear and mononuclear phagocytes, in *Handbook of Experimental Immunology: Cellular Immunology,* Vol. 2., Weir, D.M., Ed., 32.1–32.19.

White-Owen, C., Alexander, J., Sramkoski, M., and Babcock, G., 1992. Rapid whole-blood microassay using flow cytometry for measuring neutrophil phagocytosis, *J. Clin. Microbiol.,* 30, 2071–2076.

8 Oxidative Burst Assay Using Flow Cytometry

8.1 INTRODUCTION

This procedure has been developed to monitor the production of reactive oxygen species (ROS), such as hydrogen peroxide and superoxide anions, in granulocytes. The capacity to produce ROS, which is called the oxidative burst, is critical for the degradation of material by granulocytes. The oxidative burst is also implicated in inflammatory damage to tissues. In this assay, cells are exposed to a fluorescent probe to assess the production of intracellular ROS. Oxidation of the probe to a fluorescent dye, such as dichlorofluorescin diacetate (DCFH-DA) or dihydro-rhodamine 123 (DHR 123) has been used to detect ROS formation. After cellular incorporation of the probe, the cells are exposed to an activator, which stimulates the production of ROS. The fluorescence of the probe, at a time period, reveals the relative amount of intracellular ROS in a cell population. Finally, the oxidative burst is evaluated on a kinetic time basis using flow cytometry. The results should demonstrate that the production of ROS increases with time. One way to demonstrate the time effect is to make a control sample containing no activator.

The oxidative burst function for human granulocytes is performed with the following suggested material and equipment.

8.2 REAGENTS

Reagents necessary for the collection and preparation of the cell suspension (see Chapters 1 through 5):

1. Dulbecco's modified phosphate buffered saline (PBS) with no calcium and no magnesium
2. PBS with 1 g/l glucose (PBS-G)
3. 0.15 M NaCl solution, 5 mM HEPES buffer at pH 7.35 (HEPES–buffered saline)
4. DCFH-DA dissolved in ethanol 100% at a concentration of 5 mM (working solution)
5. DHR 123 dissolved in dimethyl sulfoxide (DMSO) at a concentration of 1 mM (working solution)
6. 4β-Phorbol 12β-myristate 13α-acetate (PMA) dissolved in DMSO at stock concentration of 2 mg/ml
7. N-formyl-methionyl-leucyl-phenylalanine (FMLP) dissolved in DMSO at stock concentration of 10^{-2} M

8. Lipid A, monophosphoryl (from *Salmonella minnesota* Re 595) dissolved in distilled water at stock concentration of 1 mg/ml

9. Opsonized zymosan (OZ): add 50 mg zymosan to 3 ml human plasma and 1 ml PBS; vortex and incubate for 30 min in a 37°C water bath; centrifuge at $600 \times g$ for 5 min. Wash two times with PBS and resuspend at 12.5 mg/ml OZ in 4 ml PBS

10. Sheath fluid

8.3 MATERIALS AND EQUIPMENT

Materials and equipment necessary for the collection and preparation of the cell suspension (see Chapters 1 through 5):

1. 15-ml conical polypropylene sterile tubes
2. 12×75 mm polystyrene test tubes (5 ml)
3. 0.1-ml sterile single pipetter with tips
4. Flow cytometer (equipped with an 488-nm emission argon laser)
5. Incubator set at 37°C, 5% CO_2 in humidified atmosphere for mammals and 39 to 42°C for birds
6. Water bath set at 37°C
7. 1-, 5-, and 10-ml pipettes

8.4 PROCEDURE (see Figure 8.1)

1. Adjust cell population to a concentration of 1.25×10^6 cells/ml, in PBS-G or in HEPES buffered saline according to the fluorescent probe to be used (Table 8.1), in order to obtain 1×10^6 cells in a 0.8-ml/sample.

2. Load 1×10^7 cells with either DCFH-DA or DHR 123, as indicated in Table 8.1. Incubate at 37°C in a water bath or an incubator. Do not wash cells after loading. Prepare aliquots of 1×10^6 cells (0.8 ml) for each separate treatment (including one for an unstimulated control sample).

3. Keep the loaded cells in a 37°C water bath or an incubator, protected from light before and during the flow cytometric analysis period.

4. Adjust the flow cytometer so that the fluorescence baseline of the loaded cells will fit in the first decade of the logarithmic scale of FL1. FL1 detector on flow cytometer measures DCFH-DA and DHR 123 fluorescence. Following the adjustments, record all data from the different assays, including time zero, with flow cytometer software.

5. Add only one activator per tube, according to the experimental design, and 100 µl of PBS (or PBS containing a primer to the cells). Add only 200 µl of PBS to a non-stimulated control tube (negative control tube). Vortex all tubes gently. Cells should be taken from 37°C to room temperature only for the recording of fluorescence at given time points, then returned back to 37°C. (see Table 8.2)

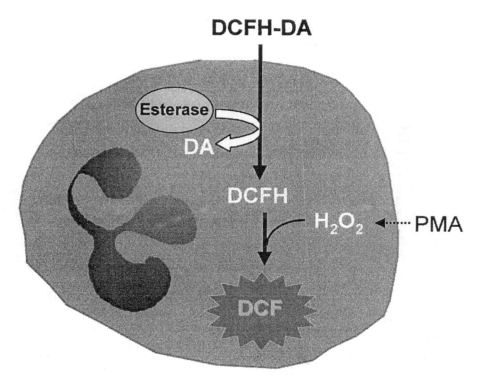

FIGURE 8.1 Principle of the oxidative burst. When cells are incubated with DCFH-DA, the probe diffuses into the cell where nonspecific esterases cleave off the acetate moiety releasing the intact substrate. Upon cell activation with stimulating agents (e.g., PMA), the hydrogen peroxide (H_2O_2) generated will oxidize the nonfluorescent probe. The resulting fluorescence measured with a fluorescein filter is proportional to the intensity of the oxidative burst.

6. After the addition of the activator to the cells, immediately begin recording the kinetics using flow cytometry. Data acquisitions should be made at times 0, 10, 20, 30, 45, and 60 min. Record 10,000 events per time point. Fewer time points can be used within the 60-min assay, if the cell number is limited.

8.5　SPECIAL RECOMMENDATIONS

1. The oxidative burst assay is primarily performed on blood granulocytes. In order to isolate granulocytes further, a forwardscatter (FSC) vs. side-scatter (SSC) gate must include this cell population in a way to exclude possible lymphocytes or monocytes since these cells might bias the results because they do not have the same oxidative burst capacity. Caution must be taken not to gate the population of granulocytes too tightly. Granulocytes might degranulate while achieving the oxidative burst and, therefore,

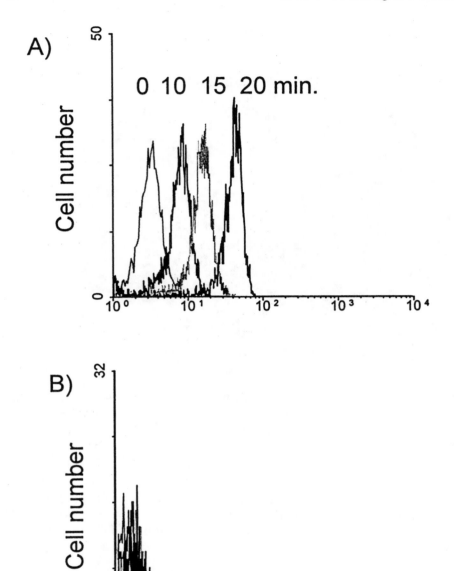

FIGURE 8.2 DCFH-DA fluorescence in human peripheral blood granulocytes. (A) Histogram showing an increase in fluorescence of human peripheral blood leukocytes; loaded with CMFDA and stimulated with PMA, over time. (B) Histogram showing the basic fluorescence of lymphocytes from the same sample. This result shows that lymphocytes do not undergo an oxidative burst.

TABLE 8.1
Fluorescent Probes

Probe	Final Concentration, μm	Working solution	Media	Incubation Time, Min.
DCFH-DA	5 μM	5 mM in ethanol[a]	PBS-G	15
DHR 123	1 μM	1 mM in DMSO[b]	Hepes-buffered saline	5

[a] Add 1 μl of DCFH-DA solution/1 ml of cell suspension.
[b] Add 1 μl of DHR 123 solution/1 ml of cell suspension.

TABLE 8.2
Cell activators

Activator	Volume Added	Final Concentration	Stock Concentration	Stock Diluted in:
PMA (FW 616.8)	100 μl	10^{-7} M	Prepare a working solution of 10^{-6} M in PBS from the original 2 mg/ml stock solution	DMSO
Lipid A[a]	100 μl	100 μg/ml	1 mg/ml	Distilled water
OZ	200 μl[b]	2.5 mg/ml	12.5 mg/ml in PBS	PBS
FMLP[c]	100 μl	1 μM	Prepare a working solution of 10^{-5} M in PBS from the original 10^{-2} M stock solution	DMSO

[a] Lipid A is a particle solution (consult special recommendations).
[b] Add only 200 μl OZ to 800 μl of cells adjusted at 1.25×10^6 cells/ml (consult special recommendations for specific fluorescence of the probe).
[c] Cells stimulated with FMLP should be primed with cytokines in order to produce a greater oxidative burst response.

reduce in FSC. Because the population might move in FSC while acquiring data, the gate should be moved to follow the movement of the cell population, or it would have to be large enough to include the population as it moves.

2. Lipid A is a particulate suspension of 100 μg/ml and could fall in the same FSC–SSC region as the granulocytes. Since lipid A is slightly fluorescent, a supplementary cytometer file containing lipid A only can be saved. The mean fluorescence of lipid A only can then be subtracted from the mean fluorescence of experimental samples containing cells and lipid A.

3. OZ is a particle solution that is slightly fluorescent. A supplementary cytometer file containing OZ only can be saved for comparisons with experimental samples containing cells and OZ.

8.6   ANALYSIS OF RESULTS

DCFH-DA and DHR 123 fluorescence is measured with an FL1 detector on the flow cytometer. By using FACScan software, FL1 histograms can be drawn for each of the experimental samples on gated (FSC–SSC) granulocytes. Histograms will show the mean fluorescence intensity (MFI) of the samples at different time points. Compare the MFI at different time points in order to evaluate the kinetics of the activation. The data can also be overlayed in an FL1 histogram.

Comparisons between MFI from experimental samples and MFI from the non-stimulated control sample will show the oxidative burst potential of different activators. A graph can be drawn using MFI vs. time points for each activator used.

8.7   SPECIES-RELATED CHANGES IN THE PROTOCOL

Species	Reagents	Materials	Procedure
Duck	1. The PMA concentration will vary with the species. It is necessary to determine the optimal PMA concentration. Concentrations ranging from 200 to 400 ng/ml were used with lesser scaup (*Aythya affinis*).	—	Same as human Keep cell at 40°C
Monkey	—	—	Same as human
Fish	1. The PMA concentration will vary with the species. It is necessary to determine the optimal PMA concentration. A concentration of 4 µg/ml was used with the rainbow trout (*Onchorhyncus mykiss*).	—	1. Use head kidney macrophages 2. Remove red blood cells
Rat and mouse	—	—	Same as human

Preparation of OZ: Proceed as described under Section 8.2, Reagents. Resuspend in PBS in order to have 5 mg OZ/ml. Fetal bovine serum can be used. Add one volume of OZ suspension to one volume of cell suspension (ratio 1:1, v/v).

Supplemented RPMI: RPMI 1640, already containing 25 mM HEPES buffer, L-glutamine, and phenol red, is supplemented with 10% (v/v) heat-inactivated fetal calf serum.

8.8   STATISTICAL TESTS

Results of many experiments can be analyzed using analysis of variance (ANOVA) on the MFI of samples for each time point, with Tukey's or Scheffé's means differentiation test. Repeated measures as a function of time can be analyzed by ANOVA.

8.9 WORKING SHEET: AQUISITION LIST

STUDY TITLE: _____   INITIALS/DATE: _____

VERIFIED/DATE: _____

DIRECTORY #/FILE NAME: _____

TUBE TAG	TIME	ACTIVATOR	MEAN FLUORESCENCE

PAGE _____ OF _____

SUGGESTED READING

Amar, M., Amit, N., Huu, T.P., Chollet-Martin, S., Labro, M.T., Gougerot-Pocidalo, M.A., and Hakim, J., 1990. Production by K562 cells of an inhibitor of adherence-related functions of human neutrophils, *J. Immunol.*, 144, 4749–4756.

Carter, W.O., Narayanan, P.K., and Robinson, J.P., 1994. Intracellular hydrogen peroxide and superoxide anion detection in endothelial cells, *J. Leukocyte Biol.*, 55, 253–258.

Elbim, C., Bailly, S., Chollet-Martin, S., Hakim, J., and Gougerot-Pocidalo, M.A., 1994. Differential priming effects of proinflammatory cytokines on human neutrophil oxidative burst in response to bacterial *N*-formyl peptides, *Infect. Immunity*, 62, 2195–2201.

Emmendorffer, A., Hecht, M., Lohmann-Matthes, M.L., and Roesler, J., 1990. A fast and easy method to determine the production of reactive oxygen intermediates by human and murine phagocytes using dihydrorhodamine 123, *J. Immunol. Methods*, 131, 269–275.

Epling, C.L., Stites, D.P., McHugh, T.M., Chong, H.O., Blackwood, L.L., and Wara, D.W., 1992. Neutrophil function screening in patients with chronic granulomatous disease by a flow cytometric method, *Cytometry*, 13, 615–620.

Gadd, M.A. and Hansbrough, J.F., 1990. Postburn suppression of murine lymphocyte and neutrophil functions is not reversed by prostaglandin blockade, *J. Surg. Res.*, 48, 84–90.

Linke, R.P., Bock, V., Valet, G., and Rothe, G., 1991. Inhibition of the oxidative burst response of *N*-formyl peptide-stimulated neutrophils by serum amyloid-A protein, *Biochem. Biophys. Res. Commun.*, 176, 1100–1105.

Macey, M.G., Sangster, J., Veys, P.A., and Newland, A.C., 1990. Flow cytometric analysis of the functional ability of neutrophils from patients with autoimmune neutropenia, *J. Microsc.*, 159, 277–283.

Métézeau, P., Ronot, X., LeNoan-Merdrignac, G., and Ratinaud, M.H., 1988. *La Cytométrie en Flux*, Vol. 1, MEDSI/McGraw-Hill, New York, 372 pp.

Ryan, T.C., Weil, G.J., Newburger, P.E., Haugland, R., and Simons, E.R., 1990. Measurement of superoxide release in the phagovacuoles of immune complex-stimulated human neutrophils, *J. Immunol. Methods*, 130, 223–233.

Samuelsson, J., Forslid, J., Hed, J., and Palmblad, J., 1994. Studies of neutrophil and monocyte oxidative responses in polycythaemia vera and related myeloproliferative disorders, *Br. J. Haematol.*, 87, 464–470.

Smith, J.A. and Weidemann, M., 1993. Further characterization of the neutrophil oxidative burst by flow cytometry, *J. Immunol. Methods*, 162, 261–268.

Tellado, J.M. and Christou, N.V., 1993. Circulating and exudative polymorphonuclear neutrophil priming and oxidative capacity in anergic surgical patients, *Arch. Surg.*, 128, 691–695.

9 Cell Cytotoxicity

9.1 INTRODUCTION

The killing of one cell type by another through direct contact constitutes a major protective mechanism against viral, certain bacterial, and perhaps cancerous cells. In this chapter, we will describe three types of target cell killing that are not major histocompatibility complex (MHC) restricted and that do not display memory or specificity:

1. Natural killer (NK) cell activity
2. Antibody-dependent cellular cytotoxicity (ADCC)
3. Lymphokine-activated cytotoxicity (LAK)

9.2 NATURAL KILLER CELL ACTIVITY

NK cells exhibit large granular lymphocyte (LGL) morphology, and they play a role in tumor resistance, host immunity to viral infections, and in the regulation of lymphoid and other hematopoietic cell populations. NK cells are present in peripheral blood and the spleen.

In this section, we describe an assay to determine the NK cell activity of human peripheral blood cells to human myeloid cell (K562) with radioactive chromium (see Figure 9.1). We also describe a radioactive free assay to evaluate NK cell activity using flow cytometry and 3,3′-dioctadecyloxacarbocyanine perchlorate (DiO) staining of YAC-1 cell line (see Figure 9.2).

9.2.1 RADIOACTIVE CHROMIUM ^{51}CR ASSAY (HUMAN OR NONHUMAN PRIMATES) AND DiO (POLAR BEAR)

9.2.1.1 Reagents

Reagents necessary for the preparation of cell suspensions (see Chapters 2 through 4).

1. RPMI 1640, containing 25 mM HEPES buffer, L-glutamine, and sodium bicarbonate
2. RPMI 1640 supplemented with 10% (v/v) heat-inactivated fetal bovine serum (completed RPMI)
3. $Na_2^{51}CrO_4$ (5 mCi/ml)
4. 5 N HCl
5. Trypan blue 0.4% (w/v) in phosphate buffer saline (PBS)
6. Supplemented RPMI with 100 U/ml penicillin and 100 µg/ml streptomycin for K562 cell line

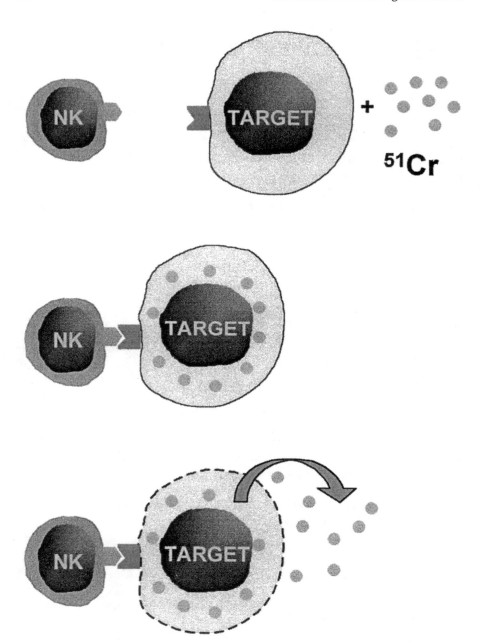

FIGURE 9.1 Principle of NK assay with ^{51}Cr. Target cells are loaded with the isotope chromium 51 (^{51}Cr). Following an incubation of 6 h with NK cells, lysed target cells will release ^{51}Cr into the medium. The level of radioactivity in the supernatant is representative of NK cytotoxicity. The principle of the LAK assay is identical, with the exception that effector cells are incubated with IL-2.

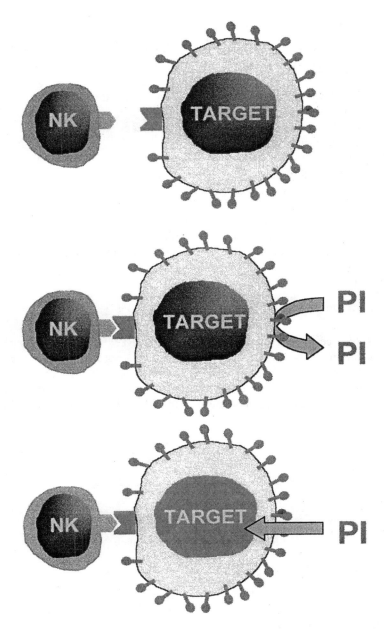

FIGURE 9.2 Principle of NK assay with DiO. Target cell membranes are stained with the DiO. DiO has an absorption and fluorescence spectra compatible with FITC. Target cells (DiO$^+$) are then incubated 2 h with effector cells in the presence of PI. When the membrane of a target cell is damaged by an NK cell, PI can no longer be excluded and the target cell is stained by PI to become DiO$^+$, PI$^+$. In contrast, target cells not affected by NK cells will exclude PI to remain DiO$^+$, PI$^-$.

9.2.1.2   Materials and Equipment

Materials and equipment necessary for the preparation of cell suspensions (see Chapters 2 through 4).

1. 96-well, round-bottom polystyrene microplate with lids
2. 15-ml conical polypropylene sterile tubes
3. 16×100 mm (15-ml) polystyrene sterile tubes with cap
4. Automatic pipetter with tips
5. 1-, 5-, and 10-ml pipettes
6. 6×50 mm flint glass disposable culture tubes
7. 50-ml sterile culture flask
8. γ counter and racks
9. Humidified incubator set at 37°C, 5% CO_2
10. Centrifuge (at room temperature) with adapter for microtiter plates
11. Two hemacytometers (one is needed to handle radioactively labeled cells)

9.2.1.3   Procedure

A. Preparation of effector cells

1. Adjust cell concentration in supplemented RPMI to the following concentrations by successive ½ dilutions and according to the respective effector:target (E:T) ratios needed as shown in Table 9.1. A minimum of 0.3 ml is needed for three wells per dilution at 100 µl/well, but there are enough lymphocytes in 20 ml of human blood to prepare 1 ml of each dilution.

TABLE 9.1
Effector and Target Cell Ratios

Ratio (E:T)	Cell Concentration (cells/ml)
100:1	1×10^7
50:1	5×10^6
25:1	2.5×10^6
12.5:1	1.25×10^6
6.25:1	6.25×10^5
3.125:1	3.125×10^5
1.56:1	1.56×10^5

B. *Preparaton of target cells*

1. Transfer K562 cells at exponential phase in a culture flask, 24 h before the experiment; therefore, be sure to add fresh medium 24 h before the experiment. Keep the cell suspension at 1×10^6 cells/ml.
2. Determine the concentration and the viability of the cells (see Chapter 5). Wash approximately 3×10^6 viable K562 cells with RPMI at room temperature in a 15-ml tube with cap. Centrifuge at $300 \times g$ for 10 min, discard supernatant, and loosen the cell pellet. Viability after washing should be greater than 95% (if not, consult special recommendations).
3. Add 100 μCi of ^{51}Cr to the cells from the initial 5 mCi stock solution. Consult half-life table provided by manufacturer for quantity of ^{51}Cr to be used. To obtain a volume, divide the optimum volume (20 μl) by the decay factor. Use safety rules for handling of radioactive material.
4. Vortex slightly and incubate for 60 min at 37°C in a humidified incubator.
5. Wash labeled K562 cells in RPMI 1640 three times. Discard supernatant using precautions for radioactive liquids.
6. Count cells in trypan blue. Resuspend at 1×10^5 cells/ml in supplemented RPMI 1640.

C. *Cytotoxic assay*

1. Run the assays in a 96-well round-bottom microplate.
2. Generally, use the experimental sequence shown in Table 9.2. Plate experimental assays in triplicate. Plate spontaneous and maximum releases in six replicates.
3. Add effector and target cells according to Table 9.3. The total volume equals 200 μl/well.

TABLE 9.2
Example of Plate Distribution of the Cells

Ratio (E:T)	Wells number
Spontaneous release	A 1–6
Maximum release	A 7–12
100:1	B 1–3
50:1	B 4–6
25:1	B 7–9
12.5:1	B 10–12
6.25:1	C 1–3
3.125:1	C 4–6
1.56:1	C 7–9

TABLE 9.3
Effector and Target Cell Ratios

Ratio (E:T)	Reagents
100:1	100 μl of effector cells (1×10^7 cells/ml) and 100 μl target cells
50:1	100 μl of effector cells (5×10^6 cells/ml) and 100 μl target cells
25:1	100 μl of effector cells (2.5×10^6 cells/ml) and 100 μl target cells
12.5:1	100 μl of effector cells (1.25×10^6 cells/ml) and 100 μl target cells
6.25:1	100 μl of effector cells (6.25×10^5 cells/ml) and 100 μl target cells
3.125:1	100 μl of effector cells (3.125×10^5 cells/ml) and 100 μl target cells
1.56:1	100 μl of effector cells (1.56×10^5 cells/ml) and 100 μl target cells
Spontaneous release	100 μl target cells and 100 μl supplemented RPMI
Maximum release	100 μl target cells and 100 μl HCl

4. Keep the rest of the labeled cells at room temperature.
5. Centrifuge plates with adapters at $300 \times g$ for 1 min.
6. Incubate for 4 h at 37°C in the incubator.
7. After the 4-h incubation period, centrifuge at $300 \times g$ for 5 min.
8. Collect 100 μl of supernatant from each tube and place this volume in separate 6×50 mm glass tubes.
9. Place 50 and 100 μl of labeled K562 cells in triplicate directly in glass tubes for estimation of total release. The 50-μl cell suspension is considered to be equal to the total release. The 100-μl cell suspension is considered to be a control for the gamma counter.
10. Place each glass tube in a sample holder and measure the release of ^{51}Cr in a γ counter.

9.2.1.4   Special Recommendations

1. *Determine the viability of K562 cells after washing:* If cell viability is low, do an isolation of cells using a density gradient ($d = 1.077$ g/ml) (consult Chapter 4; Preparation of cell suspension). Cell viability should become >95%. However, a low percentage of viability in an exponential-phase culture could be representative of a damaged cell culture.

2. *Programs for ratio calculations:* Before taking your samples to the γ counter, consult the manufacturer's operating manual. Some counters have programs that enable them to calculate the mean of the triplicate samples and to give the direct ratio cytotoxicity evaluation when the first samples are the spontaneous and maximum release samples.

3. *Safety rules for radioactive materials:* Always use gloves when handling radioactive chromium. Radioactive contaminated waste should be discarded in specific containers, according to the waste management protocols established in your facility. ^{51}Cr should be kept in a chemical hood behind lead blocks. Always check for the expired date of the radioisotope before using it.

9.2.1.5 Analysis of Results

The data are expressed as counts per minute (CPM). The average of triplicate counts is determined for each ratio and control. Then, for each E:T cell ratio, a percent cytotoxicity should be determined using the following formula:

$$\text{Percent cytotoxicity} = \frac{\text{Experimental sample (CPM)} - \text{Spontaneous release (CPM)}}{\text{Maximum release (CPM)} - \text{Spontaneous release (CPM}} \times 100$$

The percentage of total recovery can be determined by the following formula:

$$\text{Percent recovery} = \frac{\text{Total release of 50 } \mu\text{l K562 cells (CPM)}}{\text{Maximum release (CPM)}} \times 100$$

For the recovery, 50 μl of cells are needed because in experimental wells target cells are diluted 1:2 with effector cells.

The percentage of background can be determined by the following formula:

$$\text{Percent background} = \frac{\text{Spontaneous release (CPM)}}{\text{Maximum release (CPM)}} \times 100$$

Note: Percent background should not exceed 10%.

The lytic unit (LU) can be calculated with the following formula (Berger and Amos, 1979):

$$\text{Lytic unit (LU)} = \frac{10^7}{\left(\text{ET}_{20\%} \times 10^4\right)}$$

where 10^7 represents an arbitrary number of effector cells and 10^4 represents the number of target cells per well and $\text{ET}_{20\%}$ = ratio necessary to obtain 20% lysis, according to a linear regression made with the percent cytotoxicity (*y*) vs. the log ratio (*x*). The log ratio of $\text{ET}_{20\%}$ that is found with the equation of the regression is transformed to the ratio.

9.2.1.6 Expected Results and Statistical Tests

Usually, the 100:1 ratio tends to be less discriminating than the other ratios. Percentage of cytotoxicity sometimes levels off at 100:1 or higher ratios. The other ratios included in this protocol generally cover the range of the NK cytotoxic functions.

The total release is considered to be equal to the 50 μl of cell suspension. The maximum release is considered to be the supernatant of cells lysed with HCl.

The percentages of background and recovery are excluded from the statistical analysis since they represent experimental validity, not experimental control values. Experimental control groups are spontaneous (negative) and maximum (positive) releases, which are included in the percentage of NK cytotoxicity.

Statistical analysis can be directly performed on the percent cytotoxicity using analysis of variance and linear regression analysis with the different ratios. Results can also be compared using LU between different individuals.

9.2.1.7 Species-Related Changes in Protocol

Species	Reagents	Materials	Methods
Polar bear	—	—	Use YAC-1
Mice	—	—	Use YAC-1 cell line
			Effectors are spleen lymphocytes
Rat	—	—	Use YAC-1 cell line
			Effectors are spleen lymphocytes

9.2.2 FLOW CYTOMETRIC ASSAY (DiO) (POLAR BEAR)

9.2.2.1 Reagents

Reagents necessary for the collection and preparation of polar bear cell suspension:

1. RPMI 1640, containing 25 mM HEPES buffer, L-glutamine, and sodium bicarbonate
2. RPMI 1640 supplemented with 10% (v/v) heat-inactivated fetal bovine serum (supplemented RPMI)
3. DiO dissolved in dimethyl sulfoxide (DMSO) to yield 3 mM
4. Sheath fluid
5. Propidium Iodide; stock solution of 1 mg/ml in water
6. Supplemented RPMI with 100 U/ml penicillin and 100 µg/ml streptomycin for YAC-1 cell line

9.2.2.2 Materials and Equipment

Materials and equipment necessary for the collection and preparation of bear cell suspension:

1. 96-well round-bottom polystyrene microplate with lids
2. 15-ml conical polypropylene sterile tubes
3. Automatic pipetter with tips
4. 1-, 5-, and 10-ml pipettes
5. 50-ml sterile culture flask
6. Humidified incubator set at 37°C, 5% CO_2
7. Flow cytometer
8. 12 × 75 polystyrene culture tubes

TABLE 9.4
Effector and Target Cell Ratios

Ration (E:T)	Number of Effector Cells
80:1	8.0×10^5
40:1	4.0×10^5
20:1	2.0×10^5
10:1	1.0×10^5
5:1	0.5×10^5
0:1	0

9.2.2.3 Procedure

A. *Preparation of effector cells*

1. Adjust cell concentration in supplemented RPMI at 1×10^6 cells/ml and prepare the cells according to the respective E:T ratios (see Table 9.4).

B. *Preparation of target cells*

1. Pour YAC-1 cells at exponential phase (2 to 3 days culture) from 75 cm^2 flask into a centrifuge tube. Determine the concentration and the viability of the cells (see Chapter 5). Viability should be greater than 95%.
2. Transfer 1×10^7 cells to 15-ml conical tube; centrifuge at $300 \times g$ for 10 min at room temperature in order to pellet. Discard supernatant and loosen pellet by the use of a Pasteur pipette with gentle motions.
3. Place 10 µl of 3 mM DiO solution into polystyrene tubes; then forcefully add 1 ml of the YAC-1 cells to disperse the dye.
4. Incubate the tubes at 37°C in 5% CO_2 for 20 min; then wash twice with RPMI and resuspend cells in RPMI at a concentration of 1×10^6 cells/ml.

Note: DiO-stained cells can be kept for 24 h at 37°C in a 5% CO_2 incubator.

C. *Cytotoxic assay*

1. Add effector and target cells according to Table 9.5.
2. Pellet effector cells by centrifugation and resuspend in a volume of 130 µl with RPMI-stained target cells added to each tube.
3. To each tube add 130 µl of working solution of propidium iodide (PI) (100 µg/ml) followed by a centrifugation at $1000 \times g$ for 30 s to pellet the cells; then incubate for 2 h at 37°C in a 5% CO_2 incubator.
4. At the end of the incubation, suspend cells by gentle mixing just before flow cytometric analysis. The following configuration is suggested for a FACScan.

TABLE 9.5
Effector and target cell ratios

Ratio (E:T)	Number of Effector and Target Cells
80:1	800 µl effector cells + 10 µl target cells
	(8.0×10^5) (1.0×10^4)
40:1	400 µl effector cells + 10 µl target cells
	(4.0×10^5) (1.0×10^4)
20:1	200 µl effector cells + 10 µl target cells
	(2.0×10^5) (1.0×10^4)
10:1	100 µl effector cells + 10 µl target cells
	(1.0×10^5) (1.0×10^4)
5:1	50 µl effector cells + 10 µl target cells
	(0.5×10^5) (1.0×10^4)
0:1	— + 10 µl target cells
	(1.0×10^4)

FSC parameter	Linear mode (1.24)
SSC parameter	Linear mode (1.00)
FL1 parameter	Logarithmic mode detector (392)
FL2 parameter	Closed
FL3 parameter	Logarithmic mode detector (274)
No compensation	
FL1 threshold	

A total of 5000 events are saved per tube.

Note: Tubes can be kept on ice for up to 5 h before the acquisitions.

9.2.2.4   Analysis of Results

Target cell lysis is analyzed from a dot plot of FL3 vs. FL1. Live target cells are DiO^{++} and PI$^-$, while dead target cells are DiO^{++} and PI^{++}.

Using the dot plot of the control E:T ratio of 0:1, set quadrants (see Figure 9.3) as follows:

Quadrant 1 = DiO$^+$PI$^+$ represents damaged YAC-1 with partially lost membranes
Quadrant 2 = DiO^{++}PI$^+$ represents lysed YAC-1
Quadrant 3 = DiO$^+$PI$^-$ represents YAC-1 cell fragments and debris
Quadrant 4 = DiO^{++}PI$^-$ represents intact YAC-1 cells

Using the same quadrant parameters for the other samples, calculate the percentage of lysis (% lysis) with the following formula (see Figure 9.4):

$$\% \text{ lysis} = \frac{\text{quadrant 1 + quadrant 2}}{100 - \text{quadrant 3}} \times 100$$

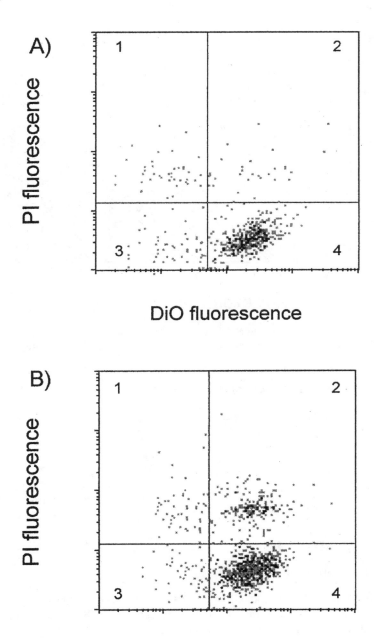

FIGURE 9.3   NK assay with polar bear peripheral blood by flow cytometry. Two-parameter cytogram (FL3 vs. FL1) of DiO/PI-stained K562 cells: (A) K562 cultured cells alone for 2 h, (B) K 562 cultured cells following a 2-h incubation with polar bear NK cells. Quadrant 1: damaged target cells. Quadrant 2: dead target cells. Quadrant 3: target cell fragments and debris. Quadrant 4: intact target cells.

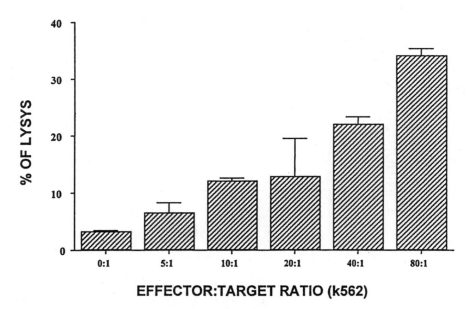

FIGURE 9.4 Typical effector:target cell ratios curve. NK activity of polar bear splenocytes evaluated by flow cytometry as the percentage of dead K-562 for each effector:target ratio.

9.2.2.5   Species-Related Changes in Protocol

Species	Reagents	Materials	Methods
Beluga	—	—	Same as polar bear
Human, nonhuman primate	—	—	Use K562 cell line; effectors are blood leukocytes
Mice	—	—	Use YAC-1 cell line; effectors are spleen lymphocytes
Rat	—	—	Use YAC-1 cell line; effectors are spleen lymphocytes

9.3   ANTIBODY-DEPENDENT CELLULAR CYTOTOXICITY

9.3.1   INTRODUCTION

Cells that perform ADCC include NK cells and cytotoxic T lymphocytes (CTL) expressing Fc receptors for immunoglobulin G (IgG) as well as macrophages and monocytes. Target cell recognition is mediated by the interaction of Fc receptors expressed on the effector cells with the Fc portion of antibody bound to the target cell.

The assay described here is the *in vitro* ADCC in mice, with chicken red blood cells (CRBC) as target cells. Effector cells are mixed with target cells (previously loaded with $Na_2{}^{51}CrO_4$) at various effector cells:target cells ratios (100:1; 50:1; 25:1). Following a 4-h incubation, the release of ^{51}Cr from lysed target cells is measured with a γ counter (see Figure 9.5).

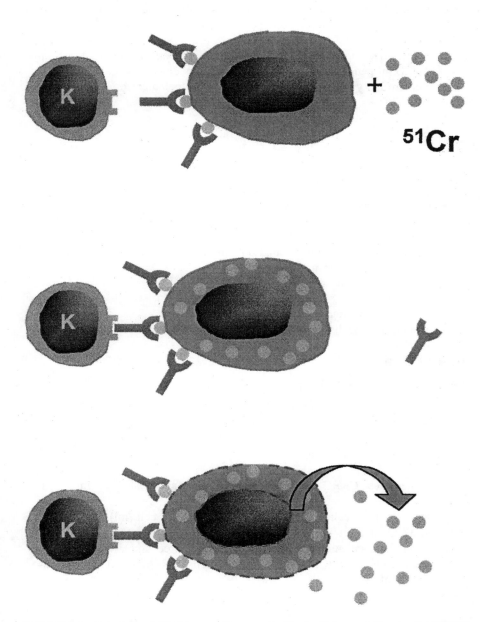

FIGURE 9.5 Principle of ADCC assay. Target cells (fresh chicken red blood cells, CRBC) are loaded with the isotope chromium 51 (^{51}Cr), followed by an incubation with an anti-CRBC serum in order to coat the target cells with antibodies. The target cells and the effector cells are then incubated together for 4 h after which the release of ^{51}Cr from lysed target cells is measured with a γ counter.

9.3.2 REAGENTS

Reagents necessary for the preparation of cell suspensions (see Chapters 2 through 4):

1. HBSS containing 10 mM HEPES
2. RPMI 1640 supplemented with 10% v/v heat-inactivated fetal calf serum, 100 U/ml penicillin, and 100 µg/ml streptomycin
3. Sodium lauryl sulfate (SLS), 2.5% (v/v) in supplemented RPMI 1640
4. $Na_2{}^{51}CrO_4$*
5. CRBC
6. Normal rabbit serum
7. Rabbit serum directed against CRBC

9.3.3 MATERIALS AND EQUIPMENT

Materials and equipment for the preparation of cell suspensions:

1. Centrifuge
2. 15-ml sterile polystyrene round-bottom tubes
3. 96-well round-bottom microtiter plate
4. 7-ml scintillation vials
5. γ counter
6. Automatic pipetter
7. Humidified incubator set at 37°C with 5% CO_2

9.3.4 PROCEDURE

A. Preparation of effector cells (E)

Prepare cell suspensions from the spleen of Balb/C mice and adjust cell suspensions at 4×10^7 cells/ml (see Chapters 3, 4, and 5).

B. Preparation of target cells (T)

1. Collect fresh CRBC in heparinized blood tubes and wash three times with HBSS.
2. Resuspend the cells in 1 ml of RPMI 1640 (2×10^7 cells/ml) and incubate at 37°C for 2 h with 100 µCi of $Na_2{}^{51}CrO_4$.
3. After the incubation period, wash the cells three times with HBSS to remove unbound ^{51}Cr and resuspend the cells in RPMI 1640 at a concentration of 4×10^5 cells/ml.

---

* Upon arrival, the activity of ^{51}Cr is 250 to 500 mCi/mg (1 mCi/ml); the half-life of ^{51}Cr is 27.7 days. If the assay is delayed after arrival of the stock solution, the volume must be increased accordingly to ensure 100 µCi of ^{51}Cr is added.

TABLE 9.6
Effector and Target Cell Ratios

Ratio E:T   REAGENTS

A. Negative Control (Normal Sera)

100:1	100 μl effector cells	+ 100 μl target cells
	$(4 \times 10^6$ cells/ml)	$(4.0 \times 10^5$ cells/ml)
50:1	100 μl effector cells	+ 100 μl target cells
	$(2 \times 10^6$ cells/ml)	$(4.0 \times 10^5$ cells/ml)
25:1	100 μl effector cells	+ 100 μl target cells
	$(1 \times 10^6$ cells/ml)	$(4.0 \times 10^5$ cells/ml)

B. Experimental Group (Antiserum-Coated CRBC)

100:1	100 μl effector cells	+ 100 μl target cells
	$(4 \times 10^6$ cells/ml)	$(4.0 \times 10^5$ cells/ml)
50:1	100 μl effector cells	+ 100 μl target cells
	$(2 \times 10^6$ cells/ml)	$(4.0 \times 10^5$ cells/ml)
25:1	100 μl effector cells	+ 100 μl target cells
	$(1 \times 10^6$ cells/ml)	$(4.0 \times 10^5$ cells/ml)

C. Control Groups

Maximum release	100 μl SLS	+ 100 μl target cells
		$(4.0 \times 10^5$ cells/ml)
Spontaneous release	100 μl supplemented	+ 100 μl target cell
	RPMI 1640	$(4.0 \times 10^5$ cells/ml)

---

4. Divide the suspension in half. To one half, add normal rabbit sera (0.2 ml/ml of cells). To the other half, add a predetermined optimal dilution of anti-CRBC rabbit serum (0.2 ml/ml of cells).
5. Incubate both suspensions at 4°C for 30 min.

C. Cytotoxicity assay

Distribute effector cells and target cells in the wells of a microplate. Suggested combinations that could be included are listed in Table 9.6. Each well should be done in triplicate.

Incubate the plate for 4 h at 37°C. Then harvest 100 μl of supernatant per well. Measure the γ activity of the released ^{51}Cr with a γ counter.

9.3.5 ANALYSIS OF RESULTS AND EXPECTED RESULTS

The average of the triplicate counts are determined in each group. The data are expressed as counts per minute (CPM). For each effector-to-target ratio, the percent cytotoxicity (cell lysis) is determined.

$$\text{Percent Cytotoxicity (PC)} = \frac{\text{Mean experimental group (CPM)} - \text{Spontaneous release (CPM)}}{\text{Maximum release (CPM)} - \text{Spontaneous release (CPM)}}$$

9.4 LYMPHOKINE-ACTIVATED KILLER CELLS

Incubation of peripheral blood lymphocytes with interleukin-2 (IL-2) results in the production of LAK cells that become cytotoxic to a wide range of neoplastic cells but not normal cells. The majority of LAK cells are derived from NK cells. In this section, we describe the lymphokine-activated cytotoxicity of peripheral blood cells from sheep to human myeloid cells K562 using radioactive chromium.

9.4.1 REAGENTS

1. RPMI 1640 supplemented with 5% fetal calf serum and 10 µg/ml of gentamycin sulfate
2. Recombinant IL-2 at 100 U/ml
3. $Na_2{}^{51}CrO_4$: Upon arrival, the activity of ^{51}Cr is 250 to 500 mCi/mg (1 mCi/ml); the half life of ^{51}Cr is 27.7 days*
4. K562 cell line
5. Triton X, 5% v/v in RPMI 1640

9.4.2 MATERIALS AND EQUIPMENT

1. 15-ml conical polystyrene tube
2. 96-well-round bottom microplate
3. Incubator set at + 37°C with 5% CO_2 and humidified atmosphere
4. γ counter
5. Automatic pipetter and tips

9.4.3 PROCEDURE

A. Preparation of effector cells

1. Resuspend isolated peripheral blood mononuclear cells in RPMI 1640 (see Chapter 2). Make dilutions in RPMI 1640 to provide concentrations of 10, 5, and 2.5 × 10^6 cells/ml. For each concentration, prepare one triplicate without and one triplicate with 20 µl of recombinant IL-2. Add volume of 100 µl of each of the dilutions to each well of a 96-well round-bottomed microtiter plate. The following suggested ratios could be included (see Table 9.7).
2. Incubate the cells at 37°C in a humidified chamber with 5% CO_2 for 24 h.
3. For each dilution, plate one triplicate with and one without IL-2.

---

* If the assay is delayed after arrival of the stock solution, the volume must be increased accordingly to ensure 200 µCi of ^{51}Cr is added.

TABLE 9.7
Effector and Target Cell Ratios

Ratio E:T   **Reagents**

A. Negative Control Group (without IL-2)

100:1	100 µl effector cells	+	100 µl target cells
	$(10.0 \times 10^6$ cells/ml)		$(1.0 \times 10^5$ cells/ml)
50:1	100 µl effector cells	+	100 µl target cells
	$(5.0 \times 10^6$ cells/ml)		$(1.0 \times 10^5$ cells/ml)
25:1	100 µl effector cells	+	100 µl target cells
	$(2.5 \times 10^6$ cells/ml)		$(1.0 \times 10^5$ cells/ml)

B. Experimental Group (with IL-2)

100:1	100 µl effector cells	+	100 µl target cells
	$(10.0 \times 10^6$ cells/ml)		$(1.0 \times 10^5$ cells/ml)
50:1	100 µl effector cells	+	100 µl target cells
	$(5.0 \times 10^6$ cells/ml)		$(1.0 \times 10^5$ cells/ml)
25:1	100 µl effector cells	+	100 µl target cells
	$(2.5 \times 10^6$ cells/ml)		$(1.0 \times 10^5$ cells/ml)

C. Control Groups

Maximum release	100 µl 5% Triton X	+	100 µl target cells
	$(1.0 \times 10^5$ cells/ml)		
Spontaneous release	100 µl supplemented	+	100 µl target cells
	RPMI 1640		$(1.0 \times 10^5$ cells/ml)

B. Preparation of target cells

1. Put K562 cells at exponential phase in a culture flask, 24 h before the experiment. Therefore, be sure to add fresh medium 24 h before the experiment. Keep the cell suspension at 1×10^6 cells/ml.

2. Determine the concentration and the viability of the cells (see Chapter 5). Wash approximately 5×10^6 K562 cells with supplemented RPMI 1640 at room temperature in a 15-ml tube with cap. Centrifuge at $300 \times g$ for 10 min, discard supernatant, and loosen the cell pellet. Viability after washing should be greater than 95% (if not, consult special recommendations, Section 9.2.1.4).

3. Resuspend the cells in 1.0 ml of supplemented RPMI 1640, and add 200 µCi of Na_2CrO_4. Consult half-life table provided by manufacturer for quantity of ^{51}Cr to be used. To obtain a volume, divide the optimum volume by the decay factor. Use safety rules for handling of radioactive material.

4. Vortex slightly and incubate for 90 min at 37°C in a humidified incubator.

5. Wash labeled K562 cells three times in supplemented RPMI 1640. Discard supernatant as a radioactive liquid.
6. Viability must be determined (see Chapter 5). Resuspend at 1×10^5 cells/ml in supplemented RPMI 1640.

C. Cytotoxic assay

1. Run the assays in the 96-well round-bottom microplate already containing the effector cells.
2. Add target cells according to Table 9.7.
3. Incubate the cell mixtures for 18 h.
4. Following the incubation period, collect 100 µl of supernatant from each well.
5. Determine the amount of ^{51}Cr released into the supernatant in a γ counter.

9.4.4   ANALYSIS OF RESULTS

9.4.4.1   Data Collection

The average of triplicate counts in each instance is determined. The data are expressed as CPM. For each ratio of effector cell to target cell, a percent cytotoxicity should be determined.

$$\text{Percent cytotoxicity} = \frac{\text{Mean sample (CPM)} - \text{Spontaneous release (CPM)}}{\text{Maximum release (CPM)} - \text{Spontaneous release (CPM)}} \times 100$$

9.4.4.2   Statistical Analysis of the Data

The percent cytotoxicity values in each group may not be normally distributed. To obtain a normal distribution, log transformation or other transformation may be required prior to use of an analysis of variance.

9.5 WORKING SHEET: ACQUISITION LIST

STUDY TITLE: _____   INITIALS/DATE: _____

VERIFIED/DATE: _____

DIRECTORY #/FILE NAME: _____

TUBE #	RATIO E:T	NOTE

PAGE ____ OF ____

SUGGESTED READING

Abbas, A.K., Lichtan, A.H., and Pober, J.S., 1991. *Cellular and Molecular Immunology,* W.B. Saunders, Philadelphia, 417 pp.

Beilin, B., Martin, F.C., Shavit, Y., Gale, R.P., and Liebeskind, J.C. Suppression of natural killer cell activity by high-dose narcotic anesthesia in rats, *Brain Behav. Immunity,* 3, 129–137.

Berger, A.E. and Amos, D.B., 1979. A comparison of antibody-dependent cellular cytotoxicity (ADCC) mediated by murine and human lymphoid cell populations, *Cell. Immunol.,* 33, 221–229.

Buttner, M., Wanke, R., and Obermann, B., 1991. Natural killer (NK) activity or porcine blood lymphocytes against allogeneic melanoma target cells, *Vet. Immunol. Immunopathol.,* 29, 89–103.

Campos, M. and Rossi, C.R., 1986. Cytotoxicity of bovine lymphocytes after treatment with lymphokines, *Am. J. Vet. Res.,* 47, 1524–1528.

Chang, L., Gusewitch, G.A., Chritton, D.B.W., Folz, J.C., Lebeck, L.K., and Nehlsen-Cannarella, S.L., 1993. Rapid flow cytometric assay for the assessment of natural killer cell activity, *J. Immunol. Methods,* 166, 45–54.

Cifone, M.G., Alesse, E., Di Eugenio, R., Napolitano, T., Morrone, S., Paolini, R., Santoni, G., and Santoni, A., 1989. *In vivo* cadmium treatment alters natural killer activity and large granular lymphocyte number in the rat, *Immunopharmacology,* 18, 149–156.

Hill, J.A., Hsia, S., Doran, D.M., and Bryans, C.I., 1986. Natural killer cell activity and antibody dependent cell-mediated cytotoxicity in preeclampsia, *J. Reprod. Immunol.,* 9, 205–212.

Hrushesky, W.J., Gruber, S.A., Sothern, R.B., Hoffman, R.A., Lakatua, D., Carlson, A., Cerra, F., and Simmons, R.L., 1988. Natural killer cell activity: age-, estrous- and circadian-stage dependence and inverse correlations with metastatic potential, *J. Natl. Cancer Inst.,* 80, 1232–1237.

Lebrec, H., Roger, K., Blot, C., Burleson, C.R., Botruon, C., and Ballardy, M., 1995. Immunotoxicological investigation using pharmaceutical drugs. *In vitro* evaluation of immune effect using rodent or human immune cells, *Toxicology,* 96, 147–156.

Racz, T., Sacks, P., Van, N.T., Taylor, D.L., Young, G., Bugis, S., Savage, H.E., and Schantz, S.P., 1990. The analysis of natural killer cell activity by flow cytometry, *Arch. Otolaryngol: Head Neck Surg.,* 116, 440–446.

Talcott, P.A., Exon, J.H., and Koller, L.D., 1984. Alteration of natural killer cell-mediated cytotoxicity in rats treated with selenium, diethylnitrosamine and ethylnitrosourea, *Cancer Lett.,* 23, 313–322.

Yagita, M., Nakajima, M., and Saksela, E., 1989. Suppression of human natural killer cell activity by amino sugars, *Cell. Immunol.,* 122, 83–95.

10 Lymphoblastic Transformation

10.1 INTRODUCTION

Immunocompetent cells require continued proliferation and differentiation for self-renewal and protection of the host against pathogens. One way to evaluate the proliferation of leukocytes is the lymphoproliferative response of B and T cells to mitogens. A mitogen is a compound that can induce mitosis in leukocytes. This procedure describes the mitogen-induced lymphoproliferative response in various vertebrate species. Phytohemagglutinin (PHA) and concanavalin A (Con A) are mitogens known to activate the proliferation of T cells, lipopolysaccharides (LPS) to induce blastogenesis in B cells, and pokeweed mitogen (PWM) to activate DNA synthesis in both T and B cells. During mitosis induced by a mitogen, the cells will incorporate the radioactive base ^{3}H-methyl thymidine. The radioactivity is measured with a β counter and the results are expressed in counts per minute (CPM) or disintegrations per minute (DPM) to give a measure of proliferation, since DPM and CPM values are proportional to the level of ^{3}H-methyl thymidine incorporated in cells (see Figure 10.1).

This mitogenic assay is described for trout head kidney blood lymphocytes with the following suggested material and equipment. Species-related differences for this test are shown in Section 10.7.

10.2 REAGENTS

1. RPMI 1640 supplemented with 4% v/v fetal calf serum (FCS), 50 U/ml penicillin, 50 µg/ml Streptomycin (supplemented RPMI)
2. Stock solution of the mitogens (PHA, Con A, LPS, PWM) prepared with supplemented RPMI 1640

   Note: For each new species, a new source of lymphocytes within one species or every new lot of mitogen, the optimal final concentration of mitogens must be determined with a standard curve. The range of concentrations for each mitogen could be as follows (see Figure 10.2):

Con A	0, 2.5, 5.0, 10.0, 20.0, 40.0 µg/ml
PHA	0, 1.0, 2.5, 5.0, 10.0, 20.0, 40.0 µg/ml
LPS	0, 5.0, 10.0, 25.0, 50.0, 100.0 µg/ml
PWM	0, 5.0, 20.0, 50.0, 100.0, 200.0, 400.0 µg/ml

3. ^{3}H-methyl thymidine: using complete RPMI 1640, prepare the working solution at 25 µCi/ml
4. Liquid scintillation cocktail
5. Distilled water

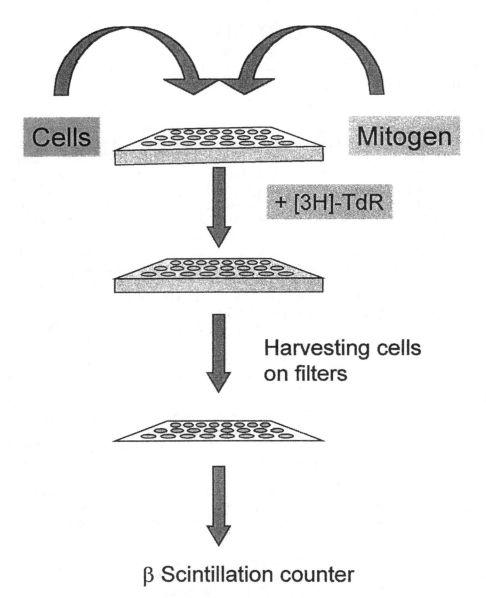

FIGURE 10.1   Principle of the mitogenic assay. Lymphocytes are incubated, with or without mitogens (e.g., Con A) and with or without xenobiotics (for toxicological studies), in a microplate. Following an appropriate incubation period, tritiated thymidine is added to the wells and the cells are incubated again to allow the incorporation of the radioactive thymidine into the DNA of dividing cells. The DNA is then captured onto glass fiber papers and the radioactivity is measured with a β scintillation counter.

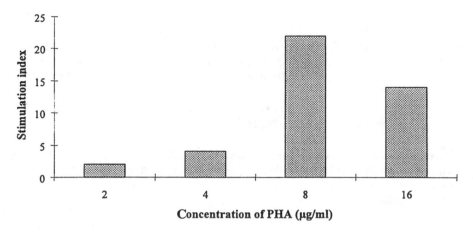

FIGURE 10.2 Standard curve with PHA in the trout model. Determination of the optimal concentration of PHA for lymphoblastic proliferation of rainbow trout head kidney leukocytes.

10.3 MATERIALS AND EQUIPMENT

1. Scintillation vials
2. Glass microfiber filters
3. Sterile 96-well flat-bottomed microplate
4. Incubator set at 20°C which provides a humidified atmosphere with 100% air environment
5. Cell harvester
6. β scintillation counter
7. Automatic pipetters
8. 12 × 75 sterile borosilicate tubes

10.4 PROCEDURE

All the steps, with the exception of the filtration, must be performed in a flow laminar hood.

1. Adjust the cell suspension to 5×10^6 cells/ml (see Chapter 4).
2. In a sterile 12 × 75 borosilicate tube, mix 0.5 ml of the cell suspension with 0.5 ml of each mitogen or with supplemented RPMI for nonstimulated control cultures. Therefore, a ½ dilution is performed for the cell suspension and the mitogens.
3. Plate each mixture (200 µl) in triplicate including the control cultures (cells incubated without mitogen). The plate must be identified.
4. Incubate the cultures at 20°C in a humidified atmosphere with 100% air environment for 48 h.
5. Add a volume of 20 µl of the ^{3}H-methyl thymidine, which contains 0.5 µCi, to each well. Then incubate the cultures in the same conditions for 18 h.

6. Using the cell harvester, aspirate the cells from each individual well onto a glass microfiber filter. Break the cells with distilled water, which leaves mainly the DNA on the filter. Then wash the wells with distilled water.
7. Remove each round-shaped filter part and place each individually in a scintillation vial correctly identified. To each vial, add a volume of 4.0 ml of scintillation cocktail.
8. Place the vials in the scintillation counter. At appropriate settings, count each vial to determine CPM and DPM.

Note: Before filtering, verify that the tubes of the cell harvester are not blocked by using a 96-well plate filled with distilled water and aspirate onto a filter. When ready to filter the plate used in the assay, wet the portion of the filter to be used with distilled water.

10.5   ANALYSIS OF RESULTS

All samples prepared in the laboratory are quenched to some degree. Therefore, in order to express the data in units that allow accurate comparison, the data must be converted from CPM to the value that actually occurred in the sample. This value is expressed as DPM.

Stimulation indexes (SI) are calculated as follows:

$$\text{Stimulation index (SI)} = \frac{\text{CPM or DPM of stimulated cultures}}{\text{CPM or DPM of unstimulated cultures}}$$

The advantage of working with stimulation indexes is that they automatically take into account the level of incorporation of unstimulated cultures and relate more to the conditions at the experiment and priming of cells (see Figure 10.3).

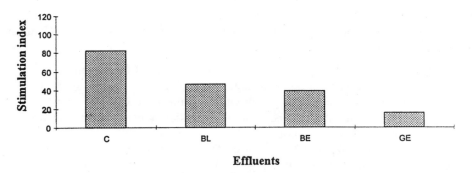

FIGURE 10.3   Effects of various pulp mill effluents on trout leukocyte mitogenesis. Effects of various pulp mill effluents, Black Liquor (BL), Bleaching Effluent (BE), and Final Effluent (FE), following a 4-day exposure period, on the proliferative response of rainbow trout head kidney leukocytes stimulated with PHA. The results are expressed as stimulation indexes ($X \pm$ SD) for eight fish per group.

The most common way to analyze the data is with an analysis of variance. However, the data obtained must be normally distributed and the homogeneity of variance must be evaluated. If the results are not normally distributed, a log transformation or other transformation may be required prior the analysis of variance.

10.6 SPECIAL RECOMMENDATIONS

1. If there is neither time nor the facilities to harvest the cells at the end of the incubation period, plates can be frozen at –20°C. This procedure will stop the lymphoproliferation without destroying the nuclei for the filtering procedure.
2. Once CPM has been obtained by counting, the DPM is calculated by the β counter with the quench curve calibration contained in the program for ^{3}H-methyl thymidine. Generally, the quench curve for ^{3}H-methyl thymidine can be calculated by the counter using radioactive thymidine standards. The quench curve can also be entered in the program if the proper coefficients of this curve are known. Calibration of the counter can be performed once a year in order to verify if a given window setting for an isotope really covers the energy spectrum for that isotope. The half-life of ^{3}H is 4492.57 days.
3. The percentage of fetal bovine serum (FBS) suggested in the table in Section 10.7 must be verified with each new lot.
4. The mitogens suggested in Section 10.7 should not be considered as exclusive. Others could be included as long as the optimal concentration is determined.
5. Always use the precautions for handling radioactive material. DPM units are often used when long-term exposure with radiomarkers is required.

10.7 SPECIES-RELATED CHANGES IN PROTOCOL

Species	Reagents	Methods
Alligator	1. Mitogens: PHA-P Con A 2. RPMI 1640 supplemented with 25 mM HEPES, 100 U/ml penicilin, 100 µg/ml streptomycin, 10% FBS, and 0.157 M (final) NaCl	1. Use peripheral blood lymphocytes 2. Prepare the cells according to Chapter 4 3. The cells are incubated at 32°C with 5% CO_2 for 96 h prior to the addition of ^{3}H-methyl thymidine and, then, for 24 h
Beluga	1. Mitogens: PHA-P Con A PWM 2. RPMI 1640 supplemented with 25 mM HEPES, 100 U/ml penicillin, 100 µg/ml streptomycin, and 10% FBS	1. Use peripheral blood lymphocytes, splenocytes, or thymocytes 2. Prepare the cells according to Chapter 4 3. The cells are incubated at 37°C with 5% CO_2 for 48 h prior to the addition of ^{3}H-methyl thymidine and, then, for 18 h
Bovine	1. Mitogens: PHA-P Con A PWM 2. RPMI 1640 supplemented with 25 mM HEPES, 100 U/ml penicillin, 100 µg/ml streptomycin, and 10% FBS	1. Use peripheral blood lymphocytes 2. Prepare the cells according to Chapter 4 3. The cells are incubated at 37°C with 5% CO_2 for 48 h prior to the addition of ^{3}H-methyl thymidine and, then, for 18 h

Species	Reagents	Methods
Horse	1. Mitogens: Con A 2. RPMI 1640 supplemented with 25 mM HEPES, 100 U/ml penicillin, 100 µg/ml streptomycin, and 10% FBS	1. Use peripheral blood lymphocytes 2. Prepare the cells according to Section 4.5.1.3B 3. The cells are incubated at 37°C with 5% CO_2 for 72 h prior to the additon of ^{3}H-methyl thymidine and, then, for 18 h

Human	1. Mitogens: PHA-P
	           Con A
	           PWM
	Note: Do not use LPS.
	2. RPMI 1640 supplemented with 25 mM HEPES, 100 U/ml penicillin, 100 µg/ml streptomycin, and 10% FBS
	3. The working dilution of ^{3}H-methyl thymidine is prepared at 100 µCi/ml and 10 ml (1 µCi) is added to each well.
		1. Use peripheral blood lymphocytes
		2. Prepare the cells according to Chapter 4 and resuspend at 1×10^6 cells/ml
		3. Plate 150 µl of cell suspension per well and add 50 µl of supplemented RPMI 1640 to nonstimulated controls
		4. The cells are incubated at 37°C with 5% CO_2 for 54 h prior to the additon of ^{3}H-methyl thymidine and, then, for 18 h

Mink	1. Mitogens: PHA-P
	           Con A
	           PWM
	2. RPMI 1640 supplemented with 25 mM HEPES, 100 U/ml penicillin, 100 µg/ml streptomycin, and 10% FBS
		1. Use peripheral blood lymphocytes
		2. Prepare the cells according to Chapter 4
		3. The cells are incubated at 37°C with 5% CO_2 for 54 h prior to the additon of ^{3}H-methyl thymidine and, then, for 18 h

Mouse	1. Mitogens: PHA-P
	           Con A
	           LPS
	2. RPMI 1640 supplemented with 25 mM HEPES, 100 U/ml penicillin, 100 µg/ml streptomycin, and 10% FBS
		1. Use splenocytes
		2. Prepare the cells according to Section 4.6.3A
		3. The cells are incubated at 37°C with 5% CO_2 for 48 h prior to the additon of ^{3}H-methyl thymidine and, then, for 18 h

Rat	1. Mitogens: PHA-P
	           Con A
	           LPS
	Note: Dextran at a final concentration of 10 µg/ml must be added to LPS.
	2. RPMI 1640 supplemented with 25 mM HEPES, 100 U/ml penicillin, 100 µg/ml streptomycin, and 10% FBS
		1. Use splenocytes
		2. Prepare the cells according to Section 4.6.3A
		3. The cells are incubated at 37°C with 5% CO_2 for 72 h prior to the additon of ^{3}H-methyl thymidine and, then, for 18 h

Seal	1. Mitogens: PHA-P
	           Con A
	           PWM
	2. RPMI 1640 supplemented with 25 mM HEPES, 100 U/ml penicillin, 100 µg/ml streptomycin, and 10% FBS
		1. Use peripheral blood lymphocytes
		2. Prepare the cells according to Chapter 4
		3. The cells are incubated at 37°C with 5% CO_2 for 48 h prior to the additon of ^{3}H-methyl thymidine and, then, for 18 h

Sheep	1. Mitogens: PHA-P Con A 2. RPMI 1640 supplemented with 25 mM HEPES, 50 U/ml penicillin, 50 µg/ml streptomycin, and 10% FBS	1. Use peripheral blood lymphocytes 2. Prepare the cells according to Chapter 4 3. The cells are incubated at 37°C with 5% CO_2 for 48 h prior to the additon of ^{3}H-methyl thymidine and, then, for 18 h

10.8 WORKING SHEETS

10.8.1 Mitogenic Assay — Working Sheet

STUDY TITLE:_____

MICROPLATE IDENTIFICATION:_____   VERIFIED/DATE: _____

	CONTROL			CON A (__ µg/ml)			PHA (__ µg/ml)			LPS (__ µg/ml)		
	1	2	3	4	5	6	7	8	9	10	11	12
A												
B												
C												
D												
E												
F												
G												
H												

A: _____
B: _____
C: _____
D: _____
E: _____
F: _____
G: _____
H: _____

Page ___ of ___

10.8.2   MITOGENIC ASSAY — FOLLOW-UP SHEET

STUDY TITLE:_____

MICROPLATE ID.:_____

VERIFIED/DATE:_____

A-   First incubation   (37°C, 5% CO_2)

      Starting time:_____

      Date: _____   Initials: _____

B- ^{3}H-Methyl thymidine

      Volume per well:_____

      Starting time: _____

      Date: _____   Initials:_____

C -   Second incubation (37°C, 5% CO_2)

      Starting time: _____

      Date: _____   Initials:_____

D -   Cell harvesting

      Starting time:_____

      Date: _____   Initials:_____

E -   Scintillation counting

      Liquid scintillation stock #: _____

      Starting time: _____

      Date: _____   Initials:_____

Page ___ of ___

10.8.3 MITOGENIC ASSAY — DPM RESULT SHEET

STUDY TITLE:_____   INITIALS/DATE:_____

VERIFIED/DATE:_____

ANIMAL	NO MITOGEN X ± S.D.	CON A ( ) X ± S.D.	PHA ( ) X ± S.D.	LPS ( ) X ± S.D.

Page ___ of ___

SUGGESTED READING

Caspi, R.R., Shahrabani, R., Kehati-Dan, T., and Avtalion, R.R., 1984. Heterogeneity of mitogen-responsive lymphocytes in carp (*Cyprinus carpio*), *Dev. Comp. Immunol.*, 8, 61–70.

Ladish, S., Ulsh, L., Gillard, B., and Wong, C., 1984. Modulation of the immune response by gangliosides. Inhibitions of adherent monocyte accessory function *in vitro*, *J. Clin. Invest.*, 76, 2074–2081.

Sharon, N., 1983. Lectin receptors as lymphocyte surface markers, *Adv. Immunol.*, 34, 213–298.

Tillit, D.E., Giesy, J.P., and Fromm, P.O., 1988. *In vitro* mitogenesis of peripheral blood lymphocytes from rainbow trout (*Salmo gairdneri*), *Comp. Biochem. Physiol.*, 89A, 25–35.

11 Determination of Antibody-Producing Cells to a Specific Antigen

11.1 INTRODUCTION

The humoral immune response is remarkable for the heterogeneity of the antibody molecules produced by B lymphocytes following an antigenic stimulation. This response requires the participation of B and T lymphocytes and macrophages; and a deficit in one or more of these cell populations will impair the humoral response. To evaluate the humoral response one can measure:

1. The level of specific antibodies secreted either by hemagglutination, which relies on the ability of antibody to cross-link sheep red blood cells (SRBC) coated with the antigen, double immunodiffusion, a simple gel assay, or by ELISA, which is a colorimetric assay to detect soluble antibodies.
2. The antibody-secreting cells following an *in vivo* or *in vitro* immunization by the plaque-forming cell (PFC) assay (the PFC assay can be performed either in liquid phase or in agar).

In this chapter, we describe an assay to evaluate the number of antibody-secreting cells in mice or rat following immunization with SRBC, a T-dependent response. Briefly, the animal is injected or a cell culture is exposed to SRBC for 4 to 5 days. Then, the spleen is removed, a splenocyte suspension is made, and the cells are exposed to the same antigen plus complement, placed in Cunningham chambers or agar plates, and left to incubate. The number of plaques formed by the lysis of SRBC by antibodies secreted and complement are counted, and the results are expressed as number of PFC/spleen or PFC/10^6 viable cells (see Figure 11.1).

11.2 LIQUID PLAQUE-FORMING CELL ASSAY IN MOUSE MODEL

11.2.1 REAGENTS

1. SRBC stored in Alsevers solution
2. Eagle's minimal essential medium supplemented with 0.2 mg/l gentamycin, sterilized by filtration; the final pH should be approximately 7.2
3. Acetic acid prepared at 1% (v/v) in a solution of 0.9% NaCl (can be made up ahead of time and stored at room temperature)
4. Guinea pig complement (GPC)

5. RPMI 1640 supplemented with 10% fetal bovine serum
6. Phosphate buffered saline (PBS) 0.05 M pH 7.4 prepared as follows:
   Solution A: 69 g $NaH_2PO_4 \cdot H_2O$ and 0.4 g thimerosal in 1 l distilled water; dissolve the thimersol first.
   Solution B: 268 g $Na_2HPO_4 \cdot 7H_2O$ and 0.4 g thimerosal in 2 l distilled water; this will need gentle heating.

   Adjust the pH of the solution B to 7.5 using solution A (about 300 ml). The remaining solution A may be stored in the cold room. Take 1.75 l of this pH 7.5 buffer (0.5 M) and add 17.5 g diEDTA and 157.5 g NaCl and heat gently until dissolved. Bring to 17.5 l. This is 0.05 M PBS.
7. Paraffin wax, 450 g
8. Petroleum jelly, 453 g
9. Deionized, distilled water
10. 95% ethanol

11.2.2 MATERIALS AND EQUIPMENT

1. Peristaltic pump
2. Cartridge filters (0.22 μ)
3. Culture dishes 60 × 15 mm
4. Culture tubes 17 × 100 mm
5. Pasteur pipettes
6. Hand tally counter
7. Glass tubes 12 × 75 mm
8. 10-ml pipettes
9. Microscope slides, plain, 75 × 25 mm
10. 1-ml tuberculin syringes for immunization
11. Double-faced masking tape
12. Eppendorf pipettes and tips
13. Crushed ice
14. 37°C hot air incubator
15. Centrifuge
16. Fluorescent lamp with dark background
17. Glass petri dish, 100 × 200 mm
18. Heating plate

11.2.3 PROCEDURE

1. On Day 0, inject mice i.p. or i.v. with 0.2 ml of 3% SRBC in PBS to stimulate antibody production. For the control group, inject mice i.p. or i.v. with PBS. Washed (3×) SRBC should be spun down at 150 × g for 10 min, then resuspended to 3% in PBS prior to immunization.
2. Before the assay date (day 5), label one set of 60 × 15 mm culture dishes and three sets of 17 × 100 mm culture tubes with the appropriate mouse number(s). Also label two sets of 12 × 75 mm culture tubes.

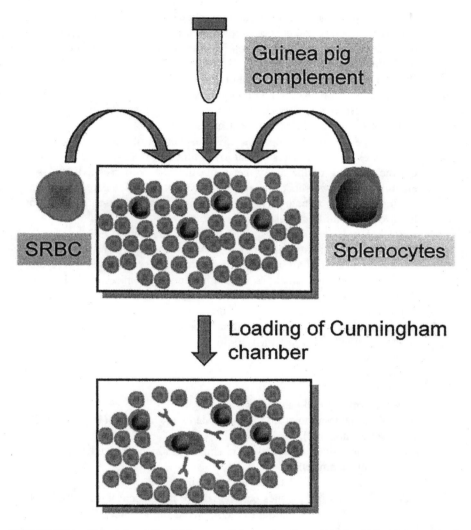

FIGURE 11.1 Principle of the PFC assay. Splenocytes from an animal already sensitized to SRBC are placed in a Cunningham chamber with SRBC and complement. Following an appropriate incubation period, plasmocytes present in the splenocytes suspension will produce and secrete antibodies specific to SRBC. The complement is then activated by the classical pathway (through binding of antibodies to surface SRBC) to lyse SRBCs, creating a clear area called plaque, around the plasmocyte, which is macroscopically countable.

3. Prepare the Cunningham chambers (counting chambers) prior to assay date as follows: clean one side of 75 × 25 mm microscope slides using 70% ethanol and a lint-free tissue. Using double-faced masking tape, cut strips approximately 3 mm wide. Onto a clean slide, carefully lay a strip of tape across the width of the slide at each end and down the center. With the cleaned side facing the tape, carefully line up and place a second

slide on top of the first. One counting slide will contain two chambers, and one counting slide is needed for each mouse. Prior to the assay, label the slides needed with the appropriate mouse number(s) and place in a 37°C incubator to prewarm.

4. Prepare a sealing mixture for counting chambers by melting paraffin wax over medium-low heat and adding to it about 10% petroleum jelly. Mix together well. If this is made up in a Pyrex petri dish, it can be covered and kept for the next assay, after it has cooled.
5. The day of the assay, prepare a 25% suspension of SRBC. Wash cells three times in PBS and resuspend to 25% in RPMI 1640.
6. On the morning of the assay, get a plastic tube with crushed ice large enough to hold two sets of labeled 17 × 100 mm tubes. Keep cells on ice at all times after mincing. Add 2 ml Eagle's minimal essential medium to all 60 × 15 mm culture dishes.
7. Immediately before making up the assay mixture, thaw GPC.
8. Euthanize the animals and prepare a cell suspension with spleen (see Chapters 3 and 4). Viability must be determined (see Chapter 5).
9. In 12 × 75 mm tubes make 2 × 10⁶/ml dilution of spleen cells in 1 ml RPMI 1640. Put petri dish of paraffin wax/petroleum jelly onto hot plate at medium-low setting to melt the mixture.
10. For each mouse, prepare assay mixture as follows:
    A. 20 µl SRBC (25% in RPMI 1640)
    B. 20 µl GPC
    C. 60 µl RPMI 1640
    Mix well, then add
    D. 100 µl Spleen cell suspension
11. Add all of the assay mixture to a preheated slide chamber (100 µl to each chamber of the slide).
12. Seal the long sides of the slide chambers with the melted mixture of paraffin/petroleum jelly.
13. Incubate all slides at 37°C for 30 to 45 min.
14. Count the number of plaques per slide under fluorescent light with a dark background. All counts should be completed immediately following the incubation period.

11.2.4 Results

The number of plaques on each counting slide should be tabulated and expressed as the number of PFC/10^6 viable spleen cells or the number of PFC per spleen.

11.3 AGAR PLAQUE-FORMING CELL ASSAY IN RAT MODEL

11.3.1 Reagents

1. SRBC stored in Alsever's solution
2. Earle's Balanced Salt Solution (EBSS)

3. GPC
4. Bacto Agar
5. DEAE dextran: stock solution is 30 mg/ml in 0.9% NaCl adjusted to pH 6.9 when dissolved

11.3.2 MATERIALS AND EQUIPMENT

1. 250-ml Pyrex flask for agar
2. 17×100 mm polypropylene tubes
3. Pasteur pipettes
4. 100×15 mm petri dish
5. 45×50 mm coverslips
6. 60×15 mm petri dish
7. 12×75 mm disposable borosilicate tubes
8. 47°C water bath
9. Centrifuge
10. 1-ml tuberculin syringes for immunization

11.3.3 PROCEDURE

1. Before the assay date, label 17×100 mm tubes with the rat identification number, fill with 6 ml EBSS within 24 h of assay, and store at 4°C; label the 100×15 mm petri dishes with animal numbers and the dilution tubes for the spleen cell suspension which will be made ($\frac{1}{50}$ and $\frac{1}{150}$ for Fischer 344 rats and $\frac{1}{50}$, $\frac{1}{200}$, and $\frac{1}{500}$ for Sprague Dawley rats).
2. Immunize rats with 2×10^6 SRBC i.v. (1 ml of 2×10^6 cells/ml). Spin down washed (3×) SRBC at $150 \times g$ for 10 min, then resuspend to 2×10^6 cells/ml in saline for immunization or in EBSS for assay.
3. Prepare a working agar solution containing 0.5% Bacto-Agar and 0.05% DEAE dextran in EBSS the morning of the assay. Prepare this by adding the agar (0.5 g/100 ml) to the buffer and dissolving while autoclaving. Then add DEAE dextran (1.6 ml stock solution/100 ml). Then dispense the warm agar in 0.35-ml aliquots into 12×75 mm disposable tubes which are held in a 47°C water bath.
4. On day 4 following immunization, sacrifice the animals, remove the spleen, and prepare a cell suspension (see Chapters 3 and 4).
5. Dilute spleen cell suspensions (see table below) in cold EBSS. Hold dilutions on ice. Viability must be determined (see Chapter 5).

Strain	IgM Assay
Fischer	$\frac{1}{50}$, $\frac{1}{150}$
Sprague-Dawley	$\frac{1}{50}$, $\frac{1}{100}$, $\frac{1}{250}$

6. For the plaque assay, each tube (in 47°C water bath) contains 0.35 ml agar/dextran; 25 µl SRBC; 0.1 ml test cells; 25 µl GPC (diluted 1:4). Follow this procedure:

   a. The SRBC should be added first, about eight tubes at a time.
   b. When the test cells are added, the tube is removed from the 47°C bath, the GPC quickly added, the tube gently vortexed, and the mixture poured into a petri dish (100 mm). The agar spot is quickly covered with a 45 × 50 mm coverslip.
   c. When the agar is solid, the plates are incubated at 37°C for 3 h. They can then be counted or stored in the refrigerator overnight.

11.3.4   RESULTS

The number of plaques on each counting slide should be tabulated and expressed as the number of $PFC/10^6$ viable spleen cells or the number of PFC/spleen (see Figure 11.2).

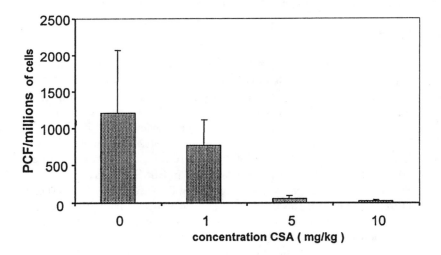

FIGURE 11.2   PFC results with rats treated with cyclosporin A using the agar method. Effect of cyclosporin A, given by gavage in rat, on the capacity of immune cells to produce an immune response against SRBC. The results are expressed in $PFC/10^6$ cells ($X \pm SD$) for 20 rats/group.

11.4   WORKING SHEET: PLAQUE-FORMING CELLS

Animal #: _____   Experiment #: _____

Type: _____

Prepared by _____

Immunization:

Animal #:	Date	Hour	Volume injected:	SRBC concentration
				ml c/ml

Cell suspension:

Spleen weight =   _____ g

Cell suspension volume =   _____ ml

Hemacytometer:

		Dead cells	Viable cells		
Counts:	1-			[Cells] =	_____ c/ml
(25 squares)	2-			% Viability =	_____ %
	Mean:				

Method used: _____
(Trypan blue or Fluorescence)

Plaque Forming Cells (PFC):   **DATE:**   _____

Petri ID:	Dilutions: *	PFC / Petri:	Mean:
	: 1/25		
	: 1/25		
	: 1/75		
	: 1/75		
	: 1/100		
	: 1/100		
	: 1/250		
	: 1/250		

* Fisher 344: 1/25, 1/75
Sprague-Dawley: 1/25, 1/100, 1/250

Results:

PFC per 10^6 viable cells: _____

PFC per spleen: _____

Prepared by: _____

SUGGESTED READING

Cunningham, A.J., 1965. A method of increased sensitivity for detecting single antibody-forming cells, *Nature,* 207, 1106–1107

Jerne, N.K., Henry, C., Nordin, A.A., Fuji, H., Koros, A.M.C., and Lefkovits, I., 1974. Plaque forming cells: methodology and theory, *Transplant. Rev.,* 18, 130–191.

Jokinen, I., Poikonen, K., and Arvilommi, H., 1985. Synthesis of human immunoglobulins *in vitro*: comparison of two assays of secreted immunoglobulin, *J. Immunoassay,* 6, 1–9.

12 Mixed Lymphocyte Reaction (MLR)

12.1 INTRODUCTION

At the very core of the immune response lies the basic phenomenon of recognition. Recognition involves histocompatibility antigens present on the surface of cells to allow the organisms to distinguish between self and nonself.

When lymphoid cells from genetically distinct organisms of the same species are mixed together in tissue culture, a reaction known as the mixed lymphocyte reaction (MLR) occurs. MLR is the anti-major histocompatibility antigens *in vitro* response and is usually interpreted as an *in vivo* analogy of graft rejection.

Lymphoid cells from each donor are stimulated by lymphoid cells from the other and undergo blast transformation. This proliferation can be observed by an increase in DNA synthesis as indicated by increased ^{3}H-methyl thymidine incorporation. In order to evaluate this mechanism adequately, the cells of one individual are pretreated with mitomycin C or irradiated to prevent their proliferation but not their surface antigens. This cell population is called the stimulating cell, and the untreated cell population is called the responding cell (see Figure 12.1).

In this chapter, the MLR assay is described for mice splenocytes.

12.2 REAGENTS

1. Hanks' balanced salt solution (HBSS) containing 10 mM HEPES
2. RPMI 1640 supplemented with 10% (v/v) heat-inactivated fetal bovine serum, 100 U/ml penicillin, 100 µg/ml streptomycin (supplemented RPMI 1640)
3. ^{3}H-methyl thymidine prepared at 25 µCi/ml in supplemented RPMI 1640
4. Liquid scintillation cocktail
5. Distilled water

12.3 MATERIALS AND EQUIPMENT

1. 15-ml sterile polystyrene round-bottom tubes
2. Sterile 96-well round-bottom microtiter plate
3. Glass microfiber filters
4. Cell harvester
5. Scintillation vials
6. β scintillation counter
7. Incubator sets at 37°C which provides a humidified atmosphere with 5% CO_2
8. Automatic pipetters and tips

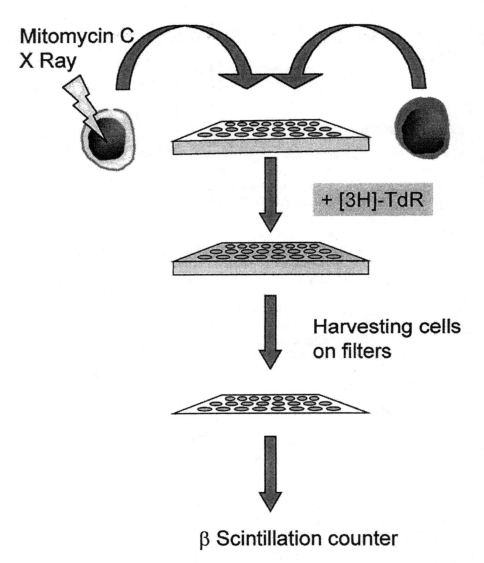

FIGURE 12.1 Principle of the mixed leukocyte culture. Mitomycin C or X-ray-treated cells (stimulating cells) from one donor are incubated with the cells of another, genetically distinct animal of the same species in a microplate. Following an appropriate incubation period, radioactive [3H]-methyl thymidine is added to the wells and the cells are incubated again to allow the incorporation of the radioactive thymidine into the DNA of dividing cells. The DNA is then captured onto glass fiber papers and the radioactivity is measured with a β scintillation counter.

12.4 PROCEDURE

1. Prepare cell suspensions from the spleen of two genetically different mice (see Chapter 3 and 4). Viability must be determined (see Chapter 5). For example, C3H(H-2^k) mice can be used as the stimulatory strain and C57Bl/6 (H-2^b) mice as the responding strain.
2. Adjust the viable cell suspensions to 5×10^6 cells/ml.
3. With an X-ray generator, irradiate the stimulating cells with 950 rad. As an alternative, the stimulating cells can be adjusted to 1×10^7 cells/ml and treated with mitomycin C (final concentration 25 µg/ml) for 20 min at 37°C in a water bath. Immediately after the incubation, 5 ml of HBSS is added to the suspension to dilute the mitomycin C. The cells are then washed three times with HBSS (wait at least 10 min between washes as mitomycin C tends to leak out of the cells and might affect the responding cells).
4. Distribute stimulating cells and responding cells in the wells of a microplate (round bottom). The following suggested combinations could be included:

   100 µl responding cells + 100 µl supplemented RPMI 1640   (Control)
   100 µl stimulating cells + 100 µl supplemented RPMI 1640   (Control)
   100 µl stimulating cells + 100 µl stimulating cells   (Negative control)
   100 µl responding cells + 100 µl stimulating cells (Experimental group)

   Plate each combination in triplicate.
5. Incubate the plates for 72 h at 37°C.
6. Add a volume of 20 µl of the ^{3}H-methyl thymidine, which contains 0.5 µCi, to each well. Then incubate the plates in the same conditions for 18 h.
7. Using the cell harvester, aspirate the cells from each individual well onto a glass microfiber filter. Break the cells with distilled water, which leaves mainly DNA on the filter. Then wash the wells with distilled water.
8. Remove each round-shaped filter part and place each individually in a scintillation vial correctly identified. To each vial, add a volume of 4.0 ml of scintillation cocktail.
9. Place the vials in the scintillation counter at appropriate settings, and count each vial to determine counts per minute (CPM) and disintegrations per minute (DPM).

Note: Before filtering, verify that the tubes of the cell harvester are not blocked by using a 96-well plate filled with distilled water and aspirate onto a filter. When ready to filter the plate used in the assay, wet the position of the filter to be used with distilled water.

12.5   ANALYSIS OF RESULTS

All samples prepared in the laboratory are quenched to some degree. Therefore, in order to express the data in units that allow accurate comparison, the data must be converted from CPM to the value that actually occurred in the sample. This value is expressed as DPM.

Stimulation indexes (SI) are calculated as follows:

$$\text{Stimulation index (SI)} = \frac{\text{CPM or DPM of stimulated cultures}}{\text{CPM or DPM of unstimulated cultures}}$$

The advantage of working with stimulation indexes is that they automatically take into account the level of incorporation of unstimulated cultures (see Figure 12.2).

The most common way to analyze the data is with an analysis of variance (ANOVA). However, the data obtained must be normally distributed and the homogeneity of variance must be evaluated. If the results are not normally distributed, a log transformation or other transformation may be required prior to the analysis of variance.

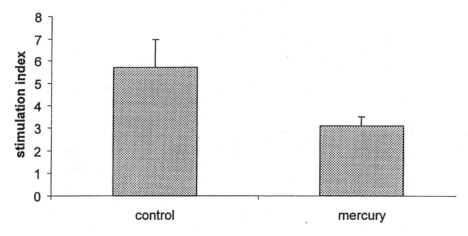

FIGURE 12.2   Mixed leukocyte culture with mouse lymphocyte exposed to methylmercury. Influence of methylmercury on the mixed lymphocyte reaction in the mouse. The exposure with mercury was performed *in vitro*. The results are presented in a stimulation index (X ± SD) from ten animals per group.

12.6   SPECIAL RECOMMENDATIONS

1. If there is neither time nor the facilities to harvest the cells at the end of the incubation period, plates can be frozen at $-20°C$. This procedure will stop the lymphoproliferation without destroying the nuclei for the filtering procedure.
2. Once CPM has been obtained by counting, the DPM is calculated by the β counter with the quench curve calibration contained in the program for ^{3}H-methyl thymidine. Generally, the quench curve for ^{3}H-methyl thymidine can be calculated by the counter using radioactive thymidine standards. The quench curve can also be entered in the program if the proper coefficients of this curve are known. Calibration of the counter can be performed once a year in order to verity if a given window setting for an isotope really covers the energy spectrum for that isotope. The half-life of ^{3}H is 4492.57 days.
3. Always use precaution when handling radioactive material.

12.7   SPECIES-RELATED CHANGES IN PROTOCOL

Species	Reagents	Materials and Equipment	Methods
Rainbow trout (*Oncorhynchus mykiss*)	RPMI 1640 + 5 mM HEPES + 5% fetal calf serum	Use head kidney cell suspensions (Chapter 3, 4, 5)	1. Irradiate the stimulating cells with 417 rad. 2. Incubate 5 days at 20°C no CO_2 3. Every 3 days, add 20 µl of RPMI 1640.
Peripheral blood lymphocytes from human or nonhuman primate	Working solution of ^{3}H-thymidine at 100 mCi/ml	—	1. Irradiate stimulating cells to 2500 rads for 5 min. 2. Adjust cell concentration to 1×10^6 cells/ml. 3. Incubate microplate at 37°C, 5% CO_2 for 120 h (5 days). 4. Add 10 µl of the 100 mCi/_l ^{3}H-thymidine solution on the 6th day. 5. Incubate for another 18 h at 37°C, 5% CO_2.

12.8   WORKING SHEETS

12.8.1   MIXED LYMPHOCYTE REACTION — WORKING SHEET

STUDY TITLE:_____

MICROPLATE IDENTIFICATION:_____   VERIFIED/DATE: _____

	1	2	3	4	5	6	7	8	9	10	11	12
A												
B												
C												
D												
E												
F												
G												
H												

A: _____
B: _____
C: _____
D: _____
E: _____
F: _____
G: _____
H: _____

12.8.2 MIXED LYMPHOCYTE REACTION — FOLLOW-UP SHEET

STUDY TITLE:_____

MICROPLATE ID.:_____

VERIFIED/DATE:_____

A - First incubation   (37°C, 5% CO_2)

      Starting time:_____
      Date: _____   Initials: _____

B - ^{3}H-Methyl·thymidine

      Volume per well:_____
      Starting time: _____
      Date: _____   Initials:_____

C - Second incubation (37°C, 5% CO_2)

      Starting time: _____
      Date: _____   Initials:_____

D - Cell harvesting

      Starting time:_____
      Date: _____   Initials:_____

E - Scintillation counting

      Liquid scintillation stock #: _____
      Starting time: _____
      Date: _____   Initials:_____

Page ___ of ___

12.8.3 Mixed Lymphocyte Reaction — DPM Result Sheet

STUDY TITLE:_____   INITIALS/DATE:_____

VERIFIED/DATE:_____

ANIMAL	X ± S.D.	X ± S.D.	X ± S.D.	X ± S.D.

SUGGESTED READING

Caspi, R.R. and Autalion, R.R., 1984. The mixed leukocyte reaction (MLR) in carp: biodirectional and unidirectional MLR responses, *Develop. Comp. Immuno. J.,* 8, 631–637.

Lubet, R.A., Brunda, M.J., Taramelli, D., Dausier, D., Nebert, D.W., and Kouri, R.E., 1984. Introduction of immunotoxicity by polycyclic hydrocarbons: role of the Ah locus, *Arch. Toxicol.,* 56, 18–24.

Thomas, P., Fugmann, R., Aranyl, C., Barbara, P., Gibbons, R., and Fenters, J., 1985. Dimethylnitrosamine exposure and immunity, *Pharmacol. Appl. Toxicol.,* 77, 219–229.

13 Intracellular Levels of Calcium Assay Using Flow Cytometry

13.1 INTRODUCTION

Calcium is an important intracellular messenger in the cell. Intracellular levels of calcium are very low compared with the extracellular environment, and calcium can be stored in intracellular compartments, such as calciosomes, the endoplasmic reticulum, and mitochondria. These intracellular pools can be mobilized by extracellular ligands upon binding of their specific membrane receptors. Extracellular calcium influx is also known to be initiated by membrane depolarization due to an external signal or by the binding of ligand to calcium channels. Certain calcium-binding fluorescent indicators, such as Fluo-3/AM (see Figure 13.1), are membrane permeable and undergo an enhancement of fluorescence on Ca^{2+} binding. Fluo-3 can therefore be used to determine differences in calcium intracellular levels using flow cytometry. Since the Fluo-3 fluorescence is dependent on the intracellular Ca^{2+} concentration, a baseline, reflecting intracellular levels of free calcium of nonactivated cells, is evaluated before any extracellular activator is incorporated into the media. Therefore, the baseline and fluorescence induced by a specific Ca^{2+} activator can be evaluated through flow cytometry.

The determination of free intracellular levels of calcium for human lymphocytes and granulocytes is performed with the following suggested material and equipment. FACStar (Becton Dickinson) results are presented in Figure 13.2.

13.2 REAGENTS

1. Density gradient (1.077 g/ml)
2. Dimethyl sulfoxide (DMSO)
3. Fluo-3/AM dissolved in DMSO at a concentration of 1 mM with 3,75% (w/v) Pluronic F-127 (keep at –20°C until use)
4. RPMI 1640, already containing 100 mg/l $Ca(NO_3)_2 \cdot 4H_2O$ (0.42 mM Ca^{2+}), L-glutamine, and phenol red
5. DNase I type IV at stock concentration of 1 mg/ml
6. Buffer A: RPMI 1640, 20 mM HEPES at pH 7.0
7. Buffer B: RPMI 1640, 20 mM HEPES, 5% (v/v) heat-inactivated fetal bovine serum at pH 7.4
8. Buffer C: RPMI 1640, 20 mM HEPES, 5% (v/v) heat-inactivated fetal bovine serum, 10 µg/ml DNase at pH 7.2

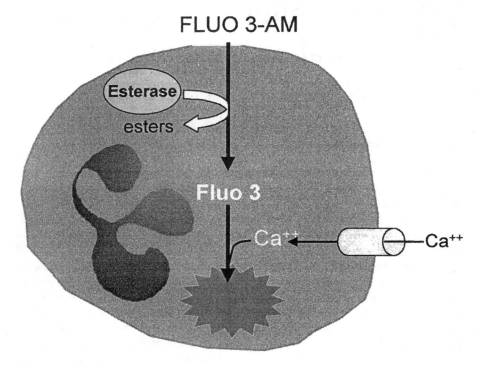

FLUO 3-AM

FIGURE 13.1 Principle of the calcium assay. Lymphocytes are loaded with Fluo-3 AM. which is de-esterified in the cytoplasm. The hydrolyzed Fluo-3 is then nonfluorescent and trapped into the cell. Following an appropriate stimulus, e.g., phytohemaglutinin (PHA), the concentration of intracellular calcium increases. The fluorescence is produced by the binding of Ca^{2+} to Fluo-3 and can be evaluated over time.

   9. Trypan blue 0.4% (w/v) in phosphate buffer solution (PBS)
 10. Sheath fluid
 11. Heparinized blood tubes
 12. Ca^{2+} activators: monoclonal antibodies, ionophores, mitogens, lympho-kines, xenobiotics

13.3   MATERIALS AND EQUIPMENT

1. 50-ml and 15-ml conical polypropylene sterile tubes
2. 2-ml borosilicate glass vials with open-top 8-mm cap and TFE-faced silicone septa
3. 0.1-ml sterile single pipetter with tips
4. 0 to 100 μl Hamilton syringe
5. 50-ml sterile culture flask

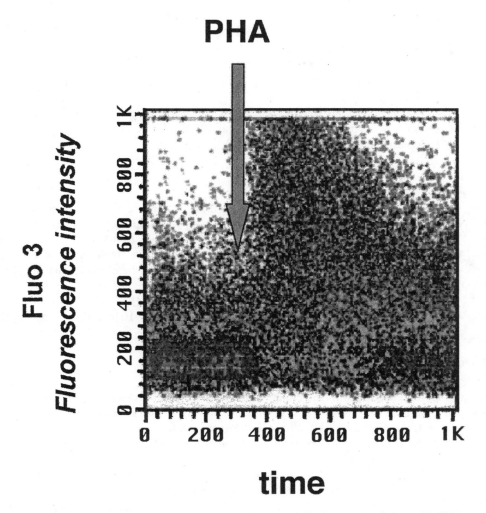

FIGURE 13.2   Calcium influx in human lymphocytes following a stimulation with PHA. Density plot representing the calcium influx induced in human lymphocytes following a stimulation with PHA.

6. Centrifuge
7. Rotating water bath set at 37°C, 50 rpm
8. Hemacytometer
9. Aluminum paper
10. pH meter
11. FACStar flow cytometer (equipped with a 488 nm emission argon laser and a miniature water bath: consult special recommendations for more details)

13.4   PROCEDURE

A. *Preparation of human lymphocytes or granulocytes*

1. The following methods are appropriate for any cell type in suspension (see Chapter 4).
2. Count cell viability (see Chapter 5). Put 5×10^6 viable cells in a 50-ml tube.
3. Wash twice in Buffer A. Centrifuge at $400 \times g$ for 10 min.
4. Resuspend at 1×10^6 cells/ml (5 ml).

B. *Loading of cells*

5. Add 1 μl/ml of 1 mM Fluo-3/AM (1 μM final concentration), or 5 μl to 5×10^6 cells (5 ml) in the 50-ml tube covered by aluminum paper (consult special recommendations for more than 5×10^6 cells to load).
6. Vortex gently and incubate at 37°C in the rotating water bath for 45 min.
7. Add an equal volume of Buffer B to the cells (5 ml). Incubate 15 min in the rotating water bath.
8. Add an equal volume of Buffer C to the cells (10 ml). Wash twice with the Buffer C (with 20 ml) and centrifuge at $400 \times g$ for 10 min. Resuspend at 5×10^5 cells/ml in Buffer C (10 ml).

C. *Flow cytometry analysis with FACStar*

9. Maintain at room temperature in the dark before flow cytometry analysis.
10. In order to have the best results, the flow cytometer is equipped with a 37°C water bath (consult special recommendations).
11. Put 0.5 ml of cell suspension in a clean borosilicate glass vial for each test performed.
12. With a first test vial, adjust flow cytometer rate with nonstimulated gated cells at 200 cells/s (consult special recommendations for fowardscatter (FSC)/sidescatter (SSC) gate). FL1 detector measures the fluorescence of Fluo-3. Adjust mean fluorescence of FL1 detector at 200 fluorescence units on a linear scale (consult special recommendations for recording).
13. With a second test vial, record cell fluorescence baseline for 8 min 32 s (512 s) with a nonstimulated control. Print the screen after the experiment is done, and clean the tubes (consult special recommendations for FACStar equipment).
14. With the third and subsequent test vials, record fluorescence baseline of nonstimulated cells for 2 min 5 s (125 s) and, then, immediately incorporate Ca^{2+} activator to the media (consult special recommendations for injection of activator). Print the screen after the experiment is done and clean the tubes.
15. For transformation of fluorescence data into intracellular concentrations (nanomolar), use one vial for stimulation with EGTA (0.5 μM final

concentration), and another vial for ionomycin (3 µM final concentration). Consult special recommendations and analysis of results. Inject after the 2 min 5 s baseline evaluation. Print the screen after the experiment is done and clean the tubes.

D. Flow cytometry analysis with another apparatus

It is possible to perform the experiment using another flow cytometer apparatus. However, the previous procedure ensures that cells are kept at 37°C during the kinetics. With a FACScan, the 12×75 mm polystyrene test tube has to be removed in order to inject the activator. The kinetics recording must be put at pause for that time. Follow the same steps for recording and analysis of data.

E. The use of other Ca^{2+} probes

There are other probes available for the determination of intracellular calcium, such as Fura-2 and INDO-1. These two calcium indicators have an excitation and emission spectra that vary with their ionic state, submitting accurate measurements by emission ratio. The utilization of such probes requires an ultraviolet excitation, but it makes dual measurements possible (e.g., calcium measurements and cell surface markers).

13.5 SPECIAL RECOMMENDATIONS

1. *Many batches of loaded cells:* If more than 5×10^6 cells are necessary for the experimental design, it is better to divide cell suspension by 5×10^6 cells and to load in separate 50-ml tubes. A separate baseline for each loading is required before the experiment. The loading of Fluo-3 is stable for 4 to 6 h, depending on the cell type and the experimental conditions. Therefore, assays should be performed immediately after the loading. If the cells are losing the Fluo-3 dye, the baseline will not be stable.

2. *FACStar special equipment:* A 37°C water bath can be fitted to the FAC-Star in order to improve the experimental conditions (consult Reference 11). The miniature water bath is fixed on the FACStar, near the flow chamber. The sample differential tube, from the main tube holder, is disconnected and replaced by another tube with a ¾-in. needle at its end. The main sample tube on the flow chamber is also disconnected (main pressure is OFF) and replaced by a new sample tube with a catheter at its end. The new differential pressure tube passes through the STANBY lock. The ¾-in. needle and the catheter (the pressure and sample tubes) go in the borosilicate vial by the septa. A plastic plier is applied on the sample tube and removed only at RUN. This ensures that no FACStar fluids would dilute the sample in the vial. However, a backwash in the sample tube can clean remaining cells. Therefore, a cleaning step is performed before every test.

3. *FSC/SSC gate and flow rate:* Before adjusting the rate with the sample differential pressure gauge, determine an FSS/SSC gate on lymphocytes with a polygon. Then, adjust the rate to 200 cells/s. A variation of ±20 cells/s is acceptable. The mean fluorescence of the FL1 detector (Fluo-3) can be adjusted with the detector voltage to fit 200 fluorescence units on the linear scale.

4. *Recording:* Record all assays with flow cytometer Lysis II software (Becton Dickinson) for 8 min 32 s (512 s at 500 ms intervals; consult time control settings on flow cytometer). Change the total recorded counts to 200,000 events and select "Direct Storage Disk." Generally, over 60,000 events are recorded. Record FSC, SSC, FL1, and time parameters. If files are to be analyzed with Chronys software (Becton Dickinson), the time parameter must not be recorded. By doing this, no Lysis II analysis can be performed after the tests.

   Mean fluorescence intensities (MFI) can also be recorded by hand during the kinetics over periods of 25 s. Furthermore, by placing a specific marker on FL1, the Lysis II software can determine the percent of cells responding to the activator. This percent of response can also be recorded by hand at different time periods. Best recording times are (1) just before the injection of the activator, (2) at maximum peak, and (3) at the end of the kinetics.

5. *Activator volume and injection:* Aspirate the exact volume of activator in the Hamilton syringe. This volume should not exceed 1/10 of the volume of the cell suspension. Add additional air in the syringe so that the activator volume will be in the middle of the syringe. This step ensures that the activator will not be immediately aspirated, as a result of pressure differences when the syringe is inserted into the vial. Clean the metal needle. The metal needle of the Hamilton syringe is then inserted into the vial. Inject the activator at 2 min 5 s (125 s).

6. *Ionomycin:* The FACStar tubes must be cleaned with distilled water for 2 min if another experiment will follow this test. For this reason, it is more appropriate to use ionomycin at the end of the experiment. If the FACStar tubes are not properly cleaned, the initial baseline will show an increase during the first 2 min 5 s. Also, a supplementary baseline can be recorded after all the experiments have been done. Under difficult experimental conditions, this supplementary control can determine if the loading of Fluo-3 was stable during the experimental time period. Validity of results can therefore be improved with this procedure.

13.6 ANALYSIS OF RESULTS

Results can be recorded with Lysis II software. Files can be analyzed through Chronys software (Becton Dickinson). MFI is calculated with cumulative fluorescence-gated data obtained over periods of 25 s. First, choose the first file to be analyzed. Then, adjust the FSC/SSC gate of Chronys. Then, change the time intervals of the mean plot graph. Intervals can be as the following: 10, 90, 110, 140, 160, 190, 210, 240, 260, 290, 310, 340, 360, 440, 460, and 512 s. Select the overlay type for the mean plot graph. Add the proper files in order to compare different stimulation, according to the experimental design. Print results (F9).

Fluorescence data can be expressed in MFI as a function of time. MFI values combine the fluorescence of approximately 5000 cells/time point.

Intracellular calcium concentrations, nanomolar (nM), can be calculated with the following formula:

$$[Ca^{2+}]_i = Kd * \frac{(F - F_{min})}{(F_{max} - F)}$$

where

Kd = represents the dissociation constant for Ca^{2+}-bound Fluo-3 and is considered to be 450 nM

F = Fluorescence of an experimental sample during the kinetics at time intervals

F_{min} = Fluorescence of the EGTA sample (negative control) at the lowest peak of the kinetics

F_{max} = Fluorescence of the ionomycin sample (positive control) at the highest peak of the kinetics

13.7   SPECIES-RELATED CHANGES IN PROTOCOL

Species	Reagents	Materials	Procedure
Mouse lymphocytes	—	—	Identical: isolate lymphocytes with density gradient.

13.8   STATISTICAL TESTS

Results of many experiments can be analyzed using analysis of variance (ANOVA), on MFI, or calcium concentrations, for each time point, with Tukey's or Scheffé's means differeniation test. Repeated measures in functions of time can be analyzed by ANOVA.

13.9   WORKING SHEET: ACQUISITION LIST

STUDY TITLE: _____   INITIALS/DATE: _____

VERIFIED/DATE: _____

DIRECTORY #/FILE NAME: _____

TUBE #	TIME	PROBE

PAGE _____ OF _____

SUGGESTED READING

Grynkiewicz, G., Poenie, M., and Tsien, R.Y., 1985. A new generation of Ca^{2+} indicators with greatly improved fluorescence properties, *J. Biol. Chem.*, 260, 3440–3450.

Kelly, K.A., 1989. Sample station modification providing on-line reagent additon and reduced sample transit time for flow cytometers, *Cytometry*, 10, 796–800.

Merritt, J.E., McCarthy, S.A., Davies, M.P., and Moores, K.E., 1990. Use of Fluo-3 to measure cytosolic Ca^{2+} in platelets and neutrophils, *Biochem. J.*, 269, 513–519.

Minta, A.K., Kao, P.Y., and Tsien, R.Y., 1989. Fluorescent indicators for cytosolic calcium based on rhodamine and fluorescein chromophores, *J. Biol. Chem.*, 264, 8171–8178.

Novak, E.J. and Rabinovitch, P.S., 1994. Improved sensitivity in flow cytometric intracellular ionized calcium measurement using fluo-3/Fura Red fluorescence ratios, *Cytometry*, 17, 135–141.

Rabinovitch, P.S. and June, C.H., 1990. Flow cytometric measurement of intracellular ionized single cells with Indo-1 and Fluo-3, in *Methods in Cell Biology*, Vol. 3, Z. Darzynkiewicz and H.A. Crissman, Eds., Academic Press, New York, 37–58.

Rijkers, G.T., Justement, L.B., Griffioen, A.W., and Cambier, J.C., 1990. Improved method for measuring intracellular Ca^{++} with fluo-3, *Cytometry*, 11, 923–927.

Sei, Y. and Arora, P., 1991. Quantitative analysis of calcium (Ca^{2+}) mobilization after stimulation with mitogens or anti-CD3 antibodies: simultaneous Fluo-3 and immunofluorescence flow cytometry, *J. Immunol. Methods*, 137, 237–244.

Vandenberghe, P.A. and Ceuppens, J.L., 1990. Flow cytometric measurement of cytoplasmic free calcium in human peripheral blood T lymphocytes with Fluo-3, a new fluorescent indicator, *J. Immunol. Methods*, 127, 197–205.

Yee, J. and Christou, N.V., 1993. Neutrophil priming by lipopolysaccharide involves heterogeneity in calcium-mediated signal transduction: studies using Fluo-3 and flow cytometry, *J. Immunol.* 150, 1988–1997.

14 Phenotyping of Blood Mononuclear Cells

14.1 INTRODUCTION

Mononuclear cell subpopulations either from peripheral blood or lymphoid organs can be characterized based upon the cell surface antigens. Fluorescent monoclonal antibodies directed against specific cell surface antigens allow for the identification of the various subpopulations of mononuclear cells using flow cytometry. Evaluation of subpopulations is useful to determine standards and help isolate pathological conditions such as cancers or immunodeficiency (see Figure 14.1).

In this chapter we will describe the phenotyping of peripheral blood mononuclear cells in the rat model.

14.2 REAGENTS

1. Dulbecco's modified phosphate buffer saline (D-PBS) at pH 7.2, without calcium and magnesium, with 0.1% (w/v) sodium azide
2. Monoclonal antibodies
3. Formalin at 0.5% (v/v) in sheath fluid
4. Sheath fluid
5. Lysing solution: The 10× stock solution is made with 89.9 g NH$_4$Cl, 10.0 g KHCO$_3$, 370.0 g tetrasodium EDTA dissolved in 1 l of distilled water. The pH is adjusted at 7.3. The working solution is obtained following ⅟₁₀ dilution of stock solution with distilled water.

14.3 MATERIALS AND EQUIPMENT

1. Automatic pipetters
2. Pipette tips
3. Timer
4. Vortex
5. Centrifuge
6. Flow cytometer
7. 12 × 75 mm polystyrene round-bottom tubes
8. 15-ml polystyrene tubes
9. Pasteur pipettes

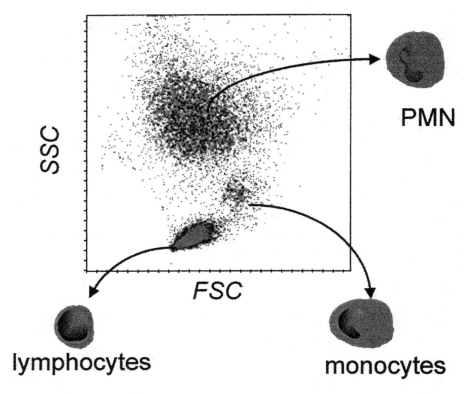

FIGURE 14.1 Scattergram of human peripheral blood. Scattergram of lysed peripheral blood. Cells are displayed according to their size (FSC; *x* axis) and complexity (SSC; *y* axis). Three populations are easily identified based on their physical properties. The lymphocytes represent an homogeneous population with a low complexity. The monocytes are larger in size and are more complex while the granulocytes are clearly more complex or granular.

14.4 PROCEDURE

1. Collect pheripheral blood in EDTA tube (see Chapter 2).
2. Distribute a volume of 100 µl of D-PBS in each 15-ml polystyrene tube.
3. Add the proper amount of isotype control reagent or specific monoclonal antibodies according to protocol. For the concentration of each isotype control reagent and monoclonal antibody, consult recommendations from the manufacturer.
4. Deliver 100 µl of whole blood to each tube and vortex gently for 2 s.
5. Incubate in the conditions recommended by the manufacturer.
6. When the incubation time is completed, lyse the red cells with prewarmed (37°C) lysing solution. To each tube, add 3.0 ml of working solution while vortexing. Then incubate all the tubes at 37°C for 5 min after the addition of the lysing solution. At the end of the incubation period, fill the tubes with D-PBS and invert three times. Recover the cells by centrifugation at

$300 \times g$ for 10 min at room temperature. Aspirate the supernatant. Approximately 50 µl of liquid should remain in the tube.

7. Wash the cells twice with D-PBS and resuspend in 0.5 ml of 0.5% formalin. Add the formalin while vortexing.

8. Keep labeled cells at 4°C in the dark. The acquisition must be done within 24 h.

9. Each acquisition should contain 10,000 events. Data on scatter parameters are acquired in list mode and data on fluorescence channels in logarithm mode. The following configuration is suggested for a FACScan (Becton Dickinson):

Forwardscatter (FSC) parameter linear mode (2.00)
Sidescatter (SSC) parameter linear mode (1.00)/detector 349
FL1 parameter logarithmic mode/detector 500
FL2 parameter logarithmic mode/detector 500

Note: When adjusting the fluorescence, the negative control evaluation should be on the far left of the scale and the positive control evaluation separated so that both peaks are not overlapping.

14.5   ANALYSIS OF RESULTS

Lymphocytes and monocytes populations are identified by their characteristic appearance on a scattergram of SSC vs. FSC (see Figure 14.2). Fluorescence signals from each population are detected on FL1 and/or FL2. The markers are set using the appropriate isotypic control to establish negative and positive populations.

The relative percentage of positive cells for each surface marker is evaluated. The absolute number could be obtained with a white blood cells count differential.

A)

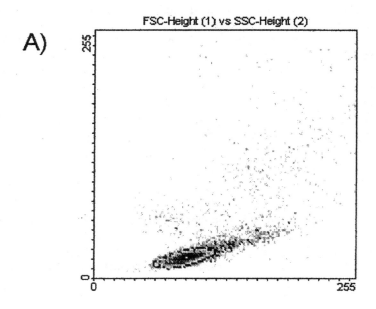

B)

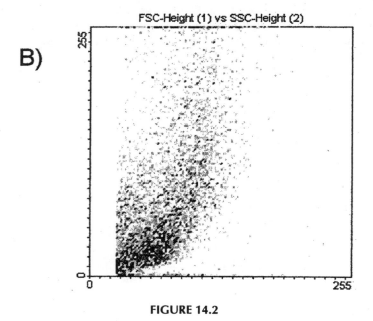

FIGURE 14.2

C)

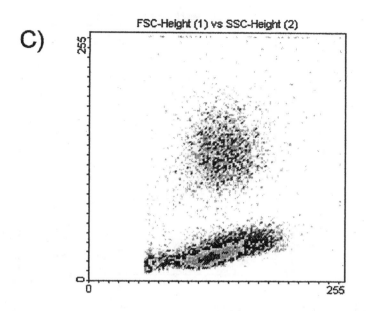

D)

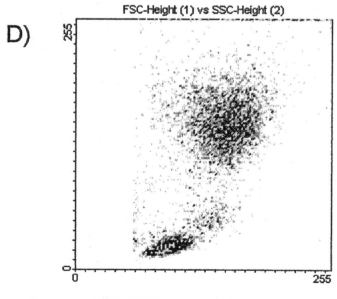

FIGURE 14.2   (continued)

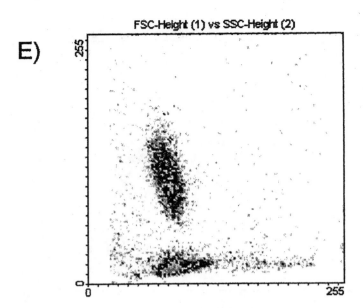

FIGURE 14.2 (continued)   Comparative scattergrams. Scattergram of the flow cytometric profile of A) alligator, B) earthworm, C) bovine, D) common seal, and E) duck peripheral blood leukocytes according to their size (FSC) and complexity (SSC).

14.6 SPECIES-RELATED CHANGES IN PROTOCOL

Species	Reagents	Materials	Methods
Sheep	PBS (0.05 M) with 0.2% gelatin GAM IgG-FITC	Mouse monoclonal directed against equine Pan T Null cells Monocytes, granulocyte CD4 CD8 Pan T B-lymphocyte	1. Buffy coat layered over a gradient with a density of 1.077 g/ml. 2. Cells are resuspended at 4.0×10^7 cells/ml in PBS gelatin and distributed in volume of 50 μl (2×10^6 cells). 3. After the incubation with each monoclonal antibody, a goat anti-mouse IgG-FITC is added. 4. The cells are then fixed with 160 μl of 2% formaldehyde.
Bovine	D-PBS with 1% bovine serum albumin and 0.1% NaN$_3$ GAM IgG-FITC	Mouse monoclonals directed against bovine MHC-class II CD2 CD4 CD8 B4 B7	1. Whole blood is diluted with an equal volume of HBSS. 2. Diluted blood is layered over a gradient with a density of 1.077 g/ml. 3. Cells are resuspended at 10×10^6 cells/ml in D-PBS-BSA and distributed in volume of 100 μl (1×10^6 cells). 4. After the first incubation with each monoclonal antibody, a goat anti-mouse IgG-FITC is added. 5. The cells are then fixed with 0.5 ml of 0.5% formalin.
Pig	As bovine	Mouse monoclonals directed against porcine. CD4 CD8 B Macrophage/monocyte	As bovine
Human	As rat	Mouse monoclonal directed against human antigens	The method is identical to the one described for the rat with the exception that the lysing solution is used at room temperature.
Cynomolgus monkey	As rat	Mouse monoclonal directed against human antigens cross-react with cynomolgus leukocytes; however, CD3 antigen is not expressed	The method is identical to the one described for the rat with the exception that the lysing solution is used at room temperature.

Note: These profiles could be modified according to the aim of the study.

14.7 IMMUNOPHENOTYPE OF RAT PERIPHERAL BLOOD LYMPHOCYTES

Surface Marker N	N	Percent Positive Lymphocytes			
		Mean	SD	Min.	Max.
OX-19+ W 3/25+ (CD4)	10	48.9	2.0	45.1	52.3
OX-19+ OX-8+ (CD8)	10	22.3	3.0	16.8	26.5
OX-6+ (RT1B)	10	21.0	1.9	17.8	24.9

Note: Antigen expression was measured on whole blood by direct immunofluorescence and analyzed on a FACScan flow cytometer. Cumulative data were obtained from ten untreated Fisher 344 female rats. Results are expressed as the mean (±SD) and as the range of relative percentage of lymphocytes expressing the indicated surface marker.

14.8 WORKING SHEETS

14.8.1 DATA LIST FOR ANTIBODIES

STUDY TITLE: _____   INITIALS/DATE: _____

VERIFIED/DATE: _____

ANTIBODY	CD	CONCENTRATION USED	VOLUME USED

PAGE _____ OF _____

14.8.2 STAINING FOLLOW-UP SHEET

STUDY TITLE: _____

INITIALS/DATE: _____

- Antibodies or D-PBS   ☐

- 100 μl of blood   ☐

- Vortex   ☐

- Incubation   Start: _____

                               Finished: _____

- Lysis of red cells   ☐

- Vortex   ☐

- Incubation   Start: _____

                               Finished: _____

- Cell washer (3 cycles) or centrifuge   ☐

- 500 μl of Formalin 0.5%   ☐

- Acquisition of data   Date: _____

                               Time: _____

PAGE _____ OF _____

14.8.3 ACQUISITION LIST

STUDY TITLE: _____   INITIALS/DATE: _____

VERIFIED/DATE: _____

DIRECTORY #/FILE NAME: _____

ANIMAL #	TUBE TAG #	CONTROL/MARKER	NOTE

PAGE _____ OF _____

SUGGESTED READING

Barclay, A.N., Birkeland, M.L., Brown, M.H., Beyers, A.D., Davis, S.J., Somaza, C., and Williams, A.F., 1993. *The Leucocyte Antigen Factsbook.* Academic Press, London.

Davis, W.C., Hamilton, M.J., Pauk, Y.H., Larsen, R.A., and Wyatt, C., 1990. Ruminant leukocytes differentiation molecules, in MHC, *Differentiation Antigens and Cytokines in Animals and Birds. Monographs in Animal Immunology,* O. Barta, Ed., Bar-Lab Inc., Blacksburg, VA, 47–70.

Davis, W.C., Marusic, S., Lewin, H., Splitter, C., Perryman, L., McGuire, T., and Gorham, J., 1987. The development and analysis of species specific and cross reactive monoclonal antibodies to leukocyte differentiation antigens and antigens of the major histocompatibility complex for use in the study of the immune system in cattle and other species, *Vet. Immunol. Immunopharmacol.,* 15, 337–376.

MacKay, C.R., Maddox, J.F., and Brandon, M.R., 1987. Lymphocyte antigens of sheep: identification and characterization using a panel of monoclonal antibodies, *Vet. Immunol. Immunopathol.,* 17, 91–102.

Moore, W. and Kautz, R., 1986. Data analysis for flow cytometry, in *The Handbook of Experiment Immunology,* 4th ed., D.M. Weir, L.A. Herzenberg, C.C. Blackwell, and L.A. Herzengerg, Eds., Blackwell Scientific, Edinburgh, 30.1–30.11.

Otterridge, P.M., Jones, W.O., and Edgar, J.A., 1988. The use of fluorescent probes and markers for sheep lymphocyte subpopulations, *Vet. Immunol. Immunopathol.,* 19, 141–151.

15 Evaluation of Intracellular Level of Thiols

15.1 INTRODUCTION

Thiols have an important role in cell proliferation, cellular transport, DNA synthesis, and cellular movements. Glutathione (GSH), a low-molecular-weight thiol, is the most abundant and important nonprotein thiol in most vertebrate cells. GSH is an important scavanger of free radicals and plays a role in the maintenance of the redox status of protein sulfhydryl groups. GSH controls the ability of lymphocytes to respond to important proliferation signals in their environment. Depletion of GSH and/or sensitive protein thiols in T lymphocytes following xenobiotic exposure may cause these cells to become nonresponsive to antigenic stimulation, resulting in immune deficiency.

The probe 5-chloromethylfluorescein diacetate (CMFDA), a colorless thiol reactive substance, freely diffuses into the cell. Then, the cytosolic esterases cleave off their acetate groups, releasing a brightly fluorescent product. This fluorescent product is then conjugated to intracellular thiols by the enzyme glutathione S-transferase. Therefore, the more thiols a cell possesses, the more fluorescent is the cell. This fluorescence can be there evaluated by flow cytometry (see Figure 15.1).

In this chapter, we describe an assay to measure the intracellular level of total thiols in mice by flow cytometry using CMFDA.

15.2 REAGENTS

1. Phosphate buffer saline (PBS) with 1 g glucose/l; the pH is adjusted at 7.4
2. CMFDA, stock solution of 5 mM with DMSO (dimethyl sulfoxide)
3. Sheath fluid

15.3 MATERIALS AND EQUIPMENT

1. 12 × 75 mm round-bottom tubes
2. Flow cytometer
3. Automatic pipetters and tips
4. Incubator set at 37°C

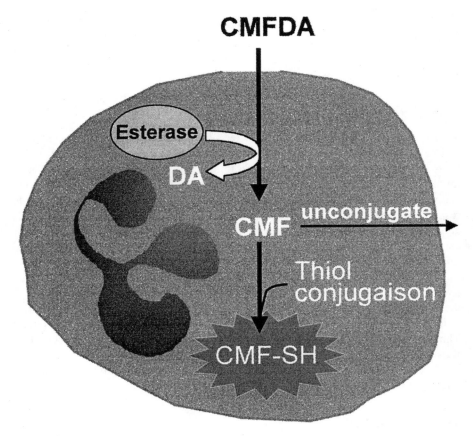

FIGURE 15.1 Principle of the thiols measurement. Cells are incubated with the nonfluorescent CMFDA, which diffuses across cell membranes where it is hydrolyzed by intracellular esterases, releasing the fluorescent reactive thiol probe. The thiol-conjugated product is trapped within the cells while the unconjugated probe diffuses into the extracellular medium.

15.4   PROCEDURE

1. Prepare a cell suspension (see Chapter 3 and 4).
2. Determine the viability of the cells (see Chapter 5) and adjust the cell suspension at 1×10^6 cells/ml in PBS-G.
3. Add 50 µM of CMFDA (10 µl of the 5 mM stock solution) and incubate 10 min at 37°C in the dark. CMFDA is retained in living cells for at least 24 h.
4. Analyze cells by flow cytometry. The following configuration is suggested for a FACScan:

   Fowardscatter (FSC) parameter   Linear mode (1.30)
   Sidescatter (SSC) parameter   Linear mode/detector at 320

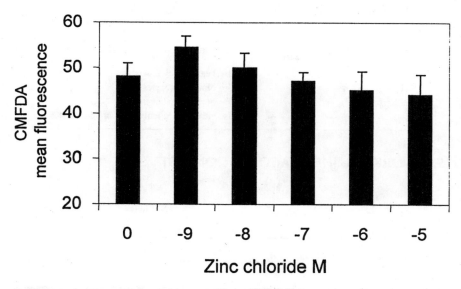

FIGURE 15.2 Influence of mercuric chloride on thiol intracellular level in mouse lymphocytes. Influence of zinc chloride on the intracellular thiol level following an *in vitro* exposure of 24 h of mouse splenocytes. The results are expressed in mean fluorescence ($X \pm SD$) from triplicates.

FL1 parameter	Logarithmic mode/detector at 382
FL2 parameter	Closed
FL3 parameter	Closed
FSC threshold	152
No compensation	
(Debris must be gated out.)	

Note:

1. A negative control group can be included using *N*-ethylmaleimide (NEM) to block intracellular thiol groups. Treat the cells with 100 µM of NEM for 10 min; then wash before staining with CMFDA.
2. When adjusting the flow cytometer one must keep in mind that an increase (stimulation) or decrease (inhibition) in the thiol levels will displace the cells in FL1.

15.5 ANALYSIS OF RESULTS

1. Before data analysis, gate the studied cell population excluding dead cells. Gate cell population on a dot plot of SSC and FL1.
2. Analyze results as mean fluorescence (histogram of FL1 from gated data) (see Figure 15.2).

15.6 SPECIES-RELATED CHANGES IN PROTOCOL

Species	Reagents	Materials	Methods
Fish	—	—	Use head kidney leukocytes
Human	—	—	Use blood leukocytes
Alligator	—	—	Use blood lymphocytes
Rat	—	—	Use either blood leukocytes or splenocytes

15.7 WORKING SHEET: ACQUISITION LIST

STUDY TITLE: _____   INITIALS/DATE: _____

VERIFIED/DATE: _____

DIRECTORY #/FILE NAME: _____

TUBE #	XENOBIOTIC	CONCENTRATION	NOTE

PAGE _____ OF _____

SUGGESTED READING

Cook, J.A. and Mitchell, J.B., 1995. Measurment of thiols in cell populations from tumor and normal tissue, *Methods Enzymol.*, 251, 203–212.
Durand, R.E. and Olive, P.L., 1983. Flow cytometry techniques for studying cellular thiols, *Radiat. Res.*, 95, 456–470.

16 Apoptosis

16.1 INTRODUCTION

Necrosis and apoptosis are two distinct modes of cell death in nucleated eukaryotic cells. Programmed cell death (apoptosis) is an important mechanism of cell selection that acts as a molecular control end point regulating physiological processes, toxicities, and diseases through cell deletion. Although the mechanism involved in apoptosis differs in various cell types, it is characterized by distinct morphological and biochemical features such as the fragmentation of DNA. Exposure to low concentrations of xenobiotics or drugs can induce apoptosis.

In this chapter we describe the evaluation of apoptotic cells by flow cytometry using Hoechst 3342 (HO3342) and propidium iodide (PI) dyes (see Figure 16.1). This method is based on the difference in the permeability of the cell to the supravital dye HO3342. Apoptotic cells incorporate HO3342 faster then viable cells. PI is added to distinguish cells that have lost membrane integrity (mainly necrotic cells and late apoptotic cells). This technique works well with thymocytes from different species.

16.2 REAGENTS

Reagents necessary for the preparation of cell suspensions (Chapter 3 through 5):

1. RPMI 1640 supplemented with 10% fetal calf serum
2. Hoechst 3342: prepare a stock solution at 50 μg/ml in distilled water; store at 4°C in the dark
3. PI: prepare a stock solution at 50 μg/ml in phosphate-buffered saline (PBS): store at 4°C in the dark
4. Dulbecco's "A" PBS
5. Dexamethasone (positive control)

16.3 MATERIALS AND EQUIPMENT

Materials and equipment necessary for the preparation of cell suspensions:

1. Flow cytometer equipped with 350 nm ultraviolet (50 mW) and 488 nm (200 mW) lasers using 575 (FL2) and 424 (FL4) band-pass filters
2. Tubes for flow cytometry

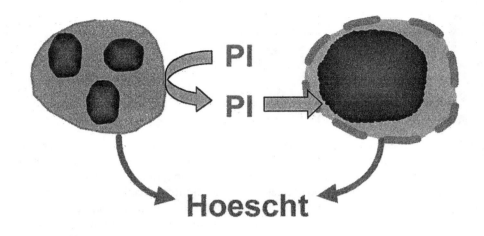

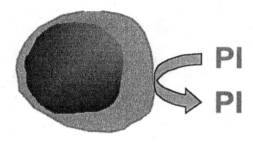

FIGURE 16.1   Principle of apoptosis. Cells are exposed to both Hoechst 33342 (excitation at 345 nm; emission at 460 nm; ultraviolet light) and PI (excitation at 488 nm; emission at 560 nm). Live cells exclude both (Ho⁻, PI⁻), whereas apoptotic cells (Ho⁺, PI⁻) only exclude PI. Dead cells (Ho⁺, PI⁺) cannot exclude either one.

16.4   PROCEDURE

1. Incubate cells in supplemented medium with or without dexamethasone (0.1 µM) for 6 h.
2. Stain cells with Hoechst 3342 (1 µg/ml) for 10 min, centrifuge, and then resuspend in PBS containing 5 µg/ml of PI.
3. Analyze cells by flow cytometry. Hoechst 3342 is excited at 360 nm (ultraviolet light) and PI at 488 nm. The following configuration is suggested for a FACS VANTAGE:

Forwardscatter (FSC) parameter	Linear mode
Sidescatter (SSC) parameter	Logarithmic mode/detector at 177
FL2 parameter	Logarithmic mode/detector at 500
FL4 parameter	Logarithmic mode/detector at 400
(Debris must be gated out.)	

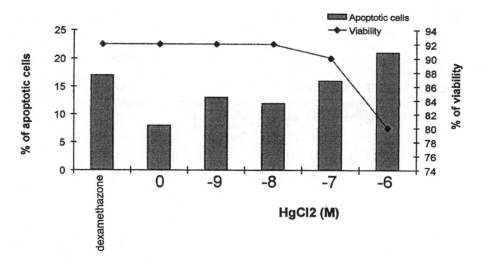

FIGURE 16.2   Influence of mercuric chloride on apoptosis in mouse thymocytes. Effect of mercuric chloride, at concentrations ranging from 10^{-6} to 10^{-9} M, on apoptosis of rat thymocytes. Dexamethazone was used as a positive control response. Viability was measured in parallel with propidium iodide.

The resultant blue vs. red fluorescence was recorded using logarithmic amplification.

16.5   ANALYSIS OF RESULTS

Necrotic cells or late apoptotic cells ($HO3342^+$ PI^+) are gated out on the basis of the uptake of PI (red fluorescence) (see Figure 16.2).

The blue fluorescent cells (normal, $HO3342^+$ PI^+ and apoptotic, $HO42^{++}$ PI^+) were gated and displayed as a cytogram of blue fluorescent intensity vs. FSC. The high-fluorescent cells were considered to be apoptotic.

SUGGESTED READING

Afnosyev, V.N., Korol, B.A., Matylevich, N.B., Pechatnikov, V.A., and Umansky, S.R., 1993. The use of flow cytometry for the investigation of cell death, *Cytometry,* 14(6), 603–609.

Cohen, J.J., 1993. Apoptosis, *Immunol. Today,* 14, 126–130.

Dolzhanskiy, A. and Basch, R.S., 1995. Flow cytometric determination of apoptosis in heterogeneous cell populations, *J. Immunol. Methods,* 180, 131–140.

Douglas, R.S., Tarshis, A.D., Pletcher, C.H., Jr., Norwell, P.C., and Moore, J.S. 1995. A simplified method for the coordinate examination of apoptosis and surface phenotype of murine lymphocytes, *J. Immunol. Methods.,* 188, 219–228.

Hamel, W., Dazin, P., and Israel, M.A., 1996. Adaptation a simple flow cytometric assay to identify different stages during apoptosis, *J. Immunol.,* 25, 173–181.

Kroemer, G., 1995. The pharmacology of T cell apoptosis, *Adv. Immunol.,* 58, 211–283.

Savill, J., Fadok, V., Henson, P., and Haslett, C., 1993. Phagocyte recognition of cells under-going apoptosis, *Immunol. Today*, 14, 131–135.

Schmid, I., Uittenbogaart, C.H., and Giorgi, J.V., 1994. Sensitive method for measuring apoptosis and cell surface phenotype in human thymocytes by flow cytometry, *Cytometry*, 15, 12–20.

Sun, X.-M., Snowden, R.T., Skilleter, D.N., Dinsdale, D., Ormerod, M.G., and Cohen, G.M., 1992. A flow cytometric method for the separation and quantitation of normal and apoptotic thymocytes, *Anal. Biochem.*, 204, 351–356.

Index

A

Acepromazine, 3
Agar plaque-forming cell assay, 90–92
Alligator
  intracellular thiol level assay, 130
  mitogenic assay, 81
Analysis of variance (ANOVA), 46
Anesthesia, 1, 3
  blood collection and, 7
  fish euthanasia, 4
Animal identification
  ear notching, 1–2
  fin clipping, 2–3
Antibody-dependent cellular cytotoxicity (ADCC), 68–72
Antibody-producing cell assay, 87–94
  agar plaque-forming cell, 90–92
  liquid plaque-forming cell, 87–90
Anticoagulants, 7
Antigen expression, cell phenotyping method, 115–125
Antigen-stimulated response
  agar plaque-forming cell assay, 90–92
  liquid plaque-forming cell assay, 87–90
  mixed lymphocyte reaction, 95–103
Apoptosis, 133–36

B

B cells
  humoral immune response, 87
  cells mitogen-induced response, 77
Bears
  blood collection, 9
  NK cell cytotoxicity assay, 57, 59, 64–68
Beluga whale
  blood collection, 9
  mitogenic assay, 81
  NK cell cytotoxicity assay, 68
Beta (β) scintillation counter, 77, 78, 81, 99
Birds
  blood collection, 10
  peritoneal exudate cell suspensions, 19–20
  white blood cell suspensions, 21–22

Blood sample collection, 7–11, See Peripheral blood collection
Bone marrow removal, 16
Bovine
  blood collection, 10–11
  mitogenic assay, 81
  mononuclear cell phenotyping, 121
  phagocytosis functional assay, 44
Brachial vein blood collection, 10

C

Calcium assay, 105–113
  alternative probes, 109
  species-specific protocol, 105
Calcium probes, 105, 109
Carbon dioxide euthanasia, 3
Cardiac puncture blood collection, 8
Cat, blood collection, 10–11
Caudal tail vein blood collection, 9
Cell cryopreservation, 33–35
Cell suspension preparation, 17–24
  avian peritoneal exudate, 19–20
  avian white blood cells, 21–22
  earthworm coelomocytes, 17–18
  fish white blood cells, 22
  lymphoid organs, 22–24
  mammalian white blood cells, 20–21
  mollusk hemocytes, 18
  rat peritoneal exudate, 18–19
Cell viability assessment, 28–31, See also Viability assessment
Centrifugation, white blood cell isolation, 20, 21
Cephalic vein blood collection, 9–10
Chicken, blood collection, 8
5-Chloromethylfluorescein diacetate (CMFDA), 127
Chromium-51 (^{51}Cr)
  antibody-dependent cytotoxicity assay, 68–72
  lymphokine-activated cytotoxicity assay, 72–74
  NK cell activity assay, 57–64
  safety rules, 62
Chronys, 110–111
CMFDA, 127
Coelomocyte extrusion, 17–18
Concanavalin A (Con A), 77